# Advanced Materials Processing and Manufacturing

This book describes the operations and industrial processes related to the production of advanced materials including ingot and powder metallurgy processing routes. It outlines the deformation processing mechanisms inducing failure at both ambient and high temperatures. Further, it embodies practical knowledge and engineering mechanisms of traditional and unorthodox material disposal approaches, concurrently with gear cutting/manufacturing, and computer numerically controlled machining. The surface fusion of metals in the production of coatings via the process of laser cladding is also covered.

Features:

- Covers novel and multi-variety techniques of materials processing and manufacturing.
- Reports on the significant variables of the processes and basic operations of advanced materials.
- Discusses fundamental and engineering machining analysis.
- Includes novel fabrication of TiAl alloys using both powder and ingot metallurgy routes.
- Enables critical thinking through technical problem solving of local service manufacturers.

This book is aimed at researchers and graduate students in materials and manufacturing engineering.

# Advanced Materials Processing and Manufacturing
*Series Editor: Kapil Gupta*

The CRC Press Series in Advanced Materials Processing and Manufacturing covers the complete spectrum of materials and manufacturing technology, including fundamental principles, theoretical background, and advancements. Considering the accelerated importance of advances for producing quality products for a wide range of applications, the titles in this series reflect the state-of-the-art in understanding and engineering the materials processing and manufacturing operations. Technological advancements for enhancement of product quality, process productivity, and sustainability are on special focus including processing for all materials and novel processes. This series aims to foster knowledge enrichment on conventional and modern machining processes. Micromanufacturing technologies such as micromachining, microforming, and microjoining, and Hybrid manufacturing, additive manufacturing, near net shape manufacturing, and ultra-precision finishing techniques are also covered.

*Advanced Materials Characterization: Basic Principles, Novel Applications, and Future Directions*
Ch Sateesh Kumar, M. Muralidhar Singh, and Ram Krishna

*Thin-Films for Machining Difficult-to-Cut Materials: Challenges, Applications, and Future Prospects*
Ch Sateesh Kumar and Filipe Daniel Fernandes

*Advanced Materials Processing and Manufacturing: Research, Technology, and Applications*
Amogelang Sylvester Bolokang and Maria Ntsoaki Mathabathe

For more information about this series, please visit: www.routledge.com/Advanced-Materials-Processing-and-Manufacturing/book-series/CRCAMPM

# Advanced Materials Processing and Manufacturing

## Research, Technology, and Applications

Amogelang Sylvester Bolokang and
Maria Ntsoaki Mathabathe

**CRC Press**
Taylor & Francis Group
Boca Raton London New York

CRC Press is an imprint of the
Taylor & Francis Group, an **informa** business

Designed cover image: (c) Shutterstock

First edition published 2024
by CRC Press
6000 Broken Sound Parkway NW, Suite 300, Boca Raton, FL 33487-2742

and by CRC Press
4 Park Square, Milton Park, Abingdon, Oxon, OX14 4RN

*CRC Press is an imprint of Taylor & Francis Group, LLC*

ISBN: 9781032411927 (hbk)
ISBN: 9781032411934 (pbk)
ISBN: 9781003356714 (ebk)

DOI: 10.1201/9781003356714

Typeset in Times
by Newgen Publishing UK

# Contents

# Figures

# Tables

# Foreword

Readers of CRC Press/Taylor and Francis advanced technology's newsletter receive monthly news on the significant developments which are continually occurring in the fields of advanced materials. They are made aware of new materials and products, advances in processing techniques, legislation, patents, and recent standards and testing procedures. They can also find commercial news, market reports, and financial information about their competitors, customers, and suppliers across the globe.

The information presented in the book is comprised of six cross-referenced chapters, each preceded by executive summary/abstract giving the authors' opinions on the principal areas of interest.

# Preface

The exploited materials of the present book are crucial in many fields of science, engineering, and technology; albeit articles reporting on 'advanced materials processing and manufacturing' are published in journals in a wide variety of fortes. Congruent to paying attention to specific advanced materials – such as, carbon fiber-reinforced plastics, TiAl-based alloys, single-crystal superalloys, and so on – countless articles are published, which are incredibly beneficial to engineers and vital to the further development of the field.

In the present book, the authors endeavour to provide synopses, case studies, and reviews on selected contents regarding advanced materials processing and manufacturing, based on research, technology, and application information to a practicable extent. Principal scientific and engineering results in advanced materials are reviewed with reference to a variety of articles published in scientific journals. The book is organized into six chapters, including the introduction in Chapter 1. Chapters 2–4 are focused on the fabrication processing of advanced materials, and Chapters 5 and 6 cover machining techniques and materials characterization, respectively. In Chapters 4 and 6 case studies, three per each chapter, are included, to facilitate scientists and forensic engineers to make neutral and objective contributions in helping to investigate and ascertain the cause of the incident or conduct research methodology.

To apprehend the science and engineering of processing advanced materials which is a wide range of fundamental knowledge of traditional and unorthodox fabrication approaches, in addition to basic knowledge of chemistry, physics, biology, and other subjects. For readers' convenience, it is recommended to consult 1) 'Laser cladding – a modern joining technique', in *Advanced Welding and Deforming*, published by Elsevier; and 2) 'Challenges in Machining of Advanced Materials', in *Advances in Sustainable Machining and Manufacturing Processes*, published by CRC Press. The inherent omniscience of a variety of issues will be expanded by the book.

It would be a great pleasure for the authors if the content of this book can provide purposeful information that can be utilized to enlighten the readers to new orientations of research.

# About the Authors

**Amogelang Sylvester Bolokang** (M-Tech) Metallurgical Engineering, (PhD) Physics, is a Principal Researcher at the Council of Scientific Industrial Research (CSIR), Adjunct Professor at the University of the Western Cape, and Associate Professor University of the Free State. He works with titanium and titanium alloys within Advanced Materials Engineering.

Sylvester's research areas are: high temperature alloys, non-ferrous and ferrous alloys (metal casting), composites, sensing, and hard materials. He works in collaboration with modelling departments to validate innovations in materials product development. His interest is in microstructure, mechanical properties, crystallography, and phase transformations of metals and alloys.

**Maria Ntsoaki Mathabathe** (PhD) Metallurgical Engineering, is a Senior Researcher based at the Council of Scientific Industrial Research (CSIR) in the Department of Materials Science and Manufacturing within Advanced Materials Engineering. She is working on advanced $\gamma$–TiAl high-temperature-based materials project, sponsored by the Department of Science and Technology (DST) and National Research Foundation (NRF), in collaboration with the CSIR.

Maria's research areas are: physical metallurgy, including oxidation and corrosion at high temperatures. In a spectrum, her research activities are a configuration of processing-structure-properties and performance relationship. Accordingly, manipulation of materials processing parameters through alloying, casting, texture analysis, laser cladding, and thermomechanical processing is evident in her research. Lastly, her interests are in the crystallography of phase domains as well as the machining and development of high-temperature advanced materials such as $\gamma$-TiAl, Ti and its alloys, NiAl based alloys, Al and their composites, and Ferrous alloys.

# 1 Overview and Prospective Challenges

## Introduction to Advanced Materials Processing and Manufacturing

*A. S. Bolokang and M. N. Mathabathe*

## 1.1 MATERIALS AND MANUFACTURING: HISTORY AND FUNDAMENTALS, GLOBAL SCENARIO

Fabrication of artefacts is dependent on raw materials, by changing their configuration, modifying and processing them into valuable finished products. At one extreme, non-conventional manufacturing techniques utilize customized machines without manual operations to accelerate large-scale manufacturing. The following is a schematic representation of the manufacturing process (see **Fig. 1.1**).

Unlike the 19th century, innovation and advancement in the steam engine and other technological applications, for example, have created a prehistoric streamlining period when industries dominate the manufacturing sector's machinery. This has resulted in the production of high-volume artefacts, and the reduction in the number of workers required to produce them.

## 1.2 CHALLENGES IN MANUFACTURING OF ENGINEERING MATERIALS AND NEED FOR ADVANCED TECHNIQUES

Producing sound products of good quality surface finish to a certain degree requires costly manufacturing processing techniques. The benefits of the added value in engineered advanced materials are a wide spread of service temperature ranges, lighter, multifunctional, and excellent shelf-life performance, which may involve expensive components such as fibres. Therefore, the produced materials require manufacturing processes that are able to accommodate the property and geometry demands of the application without degrading the fabricated artefact. In addition, manufacturing products of superior dimensional accuracy is often hindered by many factors, such as machining techniques, as a consequence of tool wear related to prolonged cutting time and high cutting force leading to increased machining costs, posing machining challenges of advanced materials [1].

DOI: 10.1201/9781003356714-1

**FIG. 1.1**   Schematic illustration for manufacturing process

## 1.3   INTRODUCTION AND CLASSIFICATION OF ADVANCED MANUFACTURING PROCESSES

Advanced manufacturing processing technologies (AMPT) have been escalating over the years because of their competitive benefits. Innumerable manufacturing industries are implementing advanced manufacturing technologies (AMT) to overcome the obstacles of maintaining quality improvement, cost depreciation, and lead time reduction, which are typically brought on by international competition, inadequate productivity growth, and quickly shifting customer tastes. Programmable computer-controlled machines are incorporated in AMT employed for the processing of large part components without incurring setup or changeover costs. These include, (1) central computer series such as computer-controlled numerical control machines (CCNCMs) or distributed numerical control machines (DNCMs); (2) computer-controlled material such as automated guided vehicle systems (AGVS), automatic storage and retrieval systems (AS/RS), and robots for welding, assembly and similar operations; and lastly (3) AMT may constitute computer-aided design (CAD). By assimilating these technologies, industries have implemented versatile manufacturing systems [2].

## 1.4   SALIENT FEATURES, APPLICATIONS, RESEARCH, AND INNOVATION

Innovation articulates socioeconomic transformation of societies; however, its negative influence such as growing incongruity amongst regions is a greater concern.

However, responsible research and innovation (RRI) is a prospective approach to mitigate these concerns, by providing sustainable regional development [3]. On one hand, extension innovation method is advantageous in both operability and universality: viz. (1) contributing to operable approaches for innovative engagements of all fields; and (2) providing quantitative and formal research in technical inventions [4] [5].

## 1.5 VARIOUS RECENT INDUSTRIAL APPLICATIONS

Industrial applications are custom tailored with the purpose of excellent cost-effectiveness and a high degree of reliability. For example, the utilization of waste-heat in industrial heat pumps applications is necessary for energy conservation.

There are various types of industrial applications such as automation fabrication, energy management and application, and process control. Tariyal and Cherin [6] outlined those applications for (1) automation fabrication involving control of machines and large data-systems, while its application controls include shipping, railways, airways, and gas and oil transportation; (2) application for process control include petrochemical, chemical, nuclear, and food industries; (3) information rates are low in automation fabrication applications when compared to fibre capacity and its optic control systems. The advantages incorporate enhanced signal quality and freedom from electromagnetic interference, and provide safety in hazardous environments, i.e. oil and gas, and remote control is achieved with ease. The following are some of the industrial applications employed where advanced materials are concerned:

- Industrial applications involving an electric motor or a generator are of paramount importance in order to connect a radial turbine to a synchronous rotational speed of the motor or generator. This integral gear technology (turbine wheel mounted directly to a gearbox) annihilates the requirement of a special driveshaft from turbine to gearbox, and hence is more maintenance friendly [7].
- Ionizing radiation in industrial applications is proliferating in recent years, which includes food irradiation, and medical and mail sterilization. At which point, material processing dosimetry systems are able to function in various conditions involving minus 60 and plus 90 °C irradiation temperatures; 0.1 and 28MeV radiation energy for electrons and photons; 1 and $10^5$ Gy the absorbed dose range, while the rate is up to $10^2$ Gys$^{-1}$ for continuous radiation, and $5\times10^7$ Gys$^{-1}$ for pulsed radiation fields [8].

## REFERENCES

[1] I. P. Tuersley, A. Jawaid, & I. R. Pashby, Review: various methods of machining advanced ceramic materials, *J. Mater. Process. Technol,* 42, (4), 377–390, 1994.

[2] T. Vipond, S. Powell, J. Daly, J. Effron, R. Kruizenga, & Y. S. Chi, Manufacturing, CFI Education Inc, 2022. https://corporatefinanceinstitute.com/resources/knowledge/other/manufacturing/

[3] R. K. Thapa, T. Iakovleva, & L. Foss, Responsible research and innovation: a systematic review of the literature and its applications to regional studies, *Eur. Plan. Stud.,* (Taylor & Francis journals), 27, (12), 2470–2490, 2019, doi: 10.1080/09654313.2019.1625871.

[4]  Y. Chun & L. Xing, Research Progress in Extension Innovation Method and its Applications, *Ind. Eng. J.*, 15, (1), 131–137, 2012.

[5]  J. V. Saraph & R. J. Sebastian, Human Resource Strategies for Effective Introduction of Advanced Manufacturing Technologies (AMT), *Prod. Invent. Manag. J.*, 33, (1), 64, 1992.

[6]  B. K. Tariyal & A. H. Cherin, Optical Fiber Communications, in *Encyclopedia of Physical Science and Technology* (Third Edition), pp. 271–294, 2003.

[7]  P. Valdimarsson, Radial inflow turbines for Organic Rankine Cycle systems, in *Organic Rankine Cycle (ORC) Power Systems*, pp. 321–334, 2017.

[8]  D. A. Schauer, A. Brodsky, & J. A. Sayeg, Radiation Dosimetry, in *Handbook of Radioactivity Analysis* (Second Edition), pp. 1165–1208, 2003.

# 2 Laser Lubrication Techniques
## Laser Processing Techniques

*A. S. Bolokang and M. N. Mathabathe*

## 2.1 INTRODUCTION

Laser-based manufacturing has been recently presented as a promising method to mitigate drawbacks encountered by conventional techniques. One of the widely used methods is the laser-based additive manufacturing (AM) to produce complex parts such as deep slots, free-form surfaces, and porous scaffolds [1]. This method is divided into three parts, viz: 1) a computer aided design (CAD) file designed to scan the object, 2) autofab software that converts the CAD data file into layers of distinct thickness of ~20-100µm, and 3) the part fabricated layer by layer using an AM machine equipped with an energy source. There are a number of energy sources including those based on electron beam, laser, and ultrasonic, for example.

Among a variety of AM technologies, the laser power-bed fusion (LPBF) method is the main contender [2, 3]. The sub-grain exclusivity cellular dislocation structure in LPBF parts is attributed to its enhanced properties. This structure is embedded in the columnar grains as various colonies with minor misorientation [4, 5]. Karthik et al. [6] investigated the role of cellular dislocation on the deformation twinning and strengthening of LPBF processed CuSn alloy. The cellular dislocation structure contributed ~45% to the yield strength of the as-built parts, while the heat-treated parts were free from the cellular Sn segregation and δ-phase [6].

Laser is defined as the stretching/expanding of light aroused by radiation discharge [7]. On one hand, lasers utilized for material processes are categorized by: wavelength (visible or ultraviolet, and infrared); output power (W, mW, kW), application (macro-processing, micromachining, etc.), operating mode (pulsed, etc.), and active medium (gas, liquid, or solid) [8]. It is important to note that the chief features of the laser beam is determined by the active mediums for material processes; the classification is shown in **Fig. 2.1**.

From an economic viewpoint, lasers contribute to the following efficacy over other processes [9]:

- Laser machining does not rely on customized environments such as a vacuum to operate, when compared to unorthodox machining processes.
- High dimensional accuracy with reduced material wastage is achieved with laser machining.

DOI: 10.1201/9781003356714-2

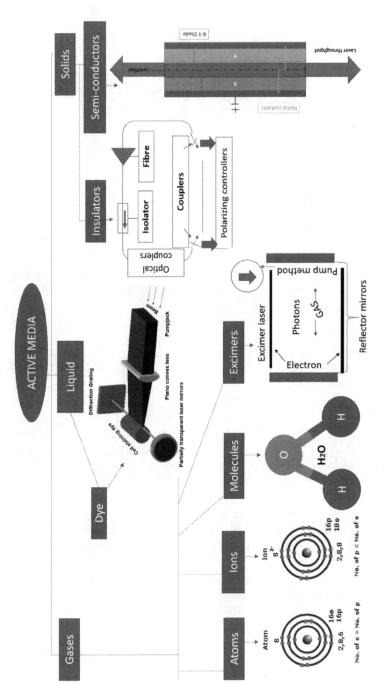

**FIG. 2.1**  Schematic representation of laser active media for material processing

**TABLE 2.1**
Economic co-relation of varying machining methods

| Machining method | | Parameters governing the economic aspects | | | | |
|---|---|---|---|---|---|---|
| | Toolings | Cost-effectiveness | Power provisions | Removal effectiveness | Tool wear | Ref |
| Conventional machining | Low | Low | Low | Very low | Low | [10–11] |
| Chemical machining | Low | Intermediate | High | Intermediate | Very low | [12–13] |
| Electrochemical machining | Intermediate | Very high | Intermediate | Low | Very low | [14–16] |
| Electrical-discharge machining | High | Intermediate | Low | High | High | [17–18] |
| Ultrasonic machining | Low | Low | Low | High | Intermediate | [19] |
| Plasma-arc machining | Low | Very low | Very low | Very low | Very low | [20–22] |
| Laser machining | Low | Intermediate | Very low | Very high | Very low | [23–25] |

- Lasers are incorporated in a computer numerical control (CNC) system to optimize machining rates.
- Recalibration and machine layoff for tool replacement are not applicable.
- Replacement costs equated with tool wear and damage are eliminated as a result of laser machining.

Further, economic comparison of laser machining with other machining processes is shown in **Table 2.1**.

This chapter reviews and addresses laser-based processing techniques and introduces the basic physical mechanisms in laser machining. Support technologies such as computers/software and post-manufacturing processes are covered. Additionally, surface improvement methods and aesthetic and property enhancement techniques are highlighted.

## 2.2   LASER MACHINING

Laser machining is a one-, two-, and three-dimensional processing technique. This method is utilized for the processing of difficult-to-cut materials such as ceramics, composites, and other hardened materials. The beneficial characteristics of laser machining apply [**9**]:

- *Laser machining is a non-contact process* - no cutting forces induced since energy transfer between the material and the laser occurs through irradiation, resulting in the absence of mechanically induced material damage, machine vibration, and tool wear.
- *Laser machining is a thermal process* - the efficacy of laser machining relies upon the thermal and optical properties, then the mechanical properties of the machined material. Thus, hard and brittle materials with acceptable thermal properties such as low thermal conductivity and diffusivity are relatively appropriate for laser machining.
- *Laser machining is a fluid process* - the laser beam can be utilized for cutting, drilling, grooving, welding, and heat-treating processes when conjoined with a multi-axis workpiece positioning system. Also, laser machining exhibits smaller kerf widths and better precision than comparable mechanical methods.

Albeit, the shortcomings associated with laser machining include [**9**]:

- *Material damage* - during the course of laser machining, the surface of the workpiece experiences high power densities that elevate the temperature of the volume to be removed to the vaporisation point. In metals, this (heat-conduction) effect generates the heat-affected zone in the surrounding area of the eroded front.
- *Low energy efficiency* - in many instances during laser machining, material volume removal ensues by melting or vaporisation. Higher energy inputs are prescribed since material removal (phase change) during laser machining occurs on an atom-by-atom basis. The establishment of 3D configuration to laser machining contributes immeasurable flexibility and an energy-efficient

**TABLE 2.2**
Laser machining parameters on varying advanced materials

| Material | Fabrication method | Parameters | Remarks | Reference |
|---|---|---|---|---|
| Al2O3 ceramic | Combination of Gelcasting and laser machining | WL:10.6, LP:30-40, FD:0.3-04, RA:1016, PA:0.025* | Al2O3 ceramic bodies with complex shapes via laser machining | [30] |
| WC | Laser machining without assist-gas | PD:0.9-10, Pf:0.3-40, LP:300 | No thermal defects observed at these set optimum parameters | [31] |
| Mesenchymal stem cell (Biomaterials) | Femtosecond laser machining | Ra:0.80-0.83, Rq: 0.98-1.05 | Good cell adhesion, proliferation and cytoskeleton properties | [32] |
| Carbon fiber reinforced plastics (CFRP) | Nd:YAG laser machining | LS:10, WL:20, PF:30, FD: 15, SV:100, WP:6, ND:150, | No matrix damage with increased strength of CFRP | [33] |
| WC-Co | | LS:0.75-6, SV:64-128, NP:8, | Optimized parameters resulted in best edge of geometry quality | [34] |
| Electrospun | Femtosecond laser machining | WL:800nm, MRR:1kHz | Effective method for preparation of electrospun biological responses | [35] |
| Cu substrate | Vibration assisted femtosecond laser machining | LS:3.5, MRR:5, WL:795nm | Controlled surface roughness and machining depth | [36] |
| Cu -Schloetter hull cell | Multi-scan underwater laser machining | LS: 165, WL:1064nm, | Copper oxidation was reduced compared to machining in air | [37] |

*WL-Wavelength (μm); LP-Laser Power(W); FD-Focus diameter (mm); RA-Resolution Accuracy (dpi); PA-Position Accuracy (mm), PD-Pulse duration (ps); PF-Pulse frequencies (MHz); Ra-Average roughness (μm); Rq-Root-mean square (μm); SV-Scanning velocity (mm/s); WP-Water pressure (MPa); ND-Nozzle diameter (μm), NP-Number of passes; MRR-Maximum repetition rate (kHz), LS-Laser scanning (μm)

material removal process. Further, advancement in industrial lasers may elevate
the proficiency of converting electrical energy to beam energy.

For mechanical material removal, lasers are preferred in a variety of engineering
applications due to their favourable salient features in thermal treatment processes.
The use of lasers can replace mechanical material removal methods in various engineering applications. According to Samant & Dahotre [26], these include: 1) Non-
contact process-energy transfer process eliminates tool wear, machine vibration, and
cutting forces; 2) Flexible process-lasers utilized for welding, drilling, grooving,
cutting, and heat-treatment; 3) Thermal process – this makes hard and brittle materials
like ceramics with low thermal conductivity/diffusivity easier for machining. On the
other hand, some setbacks encountered by the lasers include cost-ineffectiveness,
thickness limitations, and material restrictions [27].

Ultimate energy lasers are an incipient industrial tool to produce complex shapes
on brittle and hard ceramic structures like alumina; refer to **Table 2.2**. Recently,
combined computational and experimental methods have been explored to determine
the mechanism of laser machining on $Al_2O_3$ and its corresponding effects on surface topography. The work by Vora et al. [28] studied this model that incorporated
phase-change kinetics, temperature-depended material properties, and empirical
boundary conditions. The best results were shown for single-pulse one-dimensional
laser machining, with less material removal and good surface finish, with the help of
a computational model [29].

## 2.3  LASER WELDING

Laser welding is a joining technology for advanced metallic alloys described by
fusion and heat-affected zones, coupled with non-existent distortions. Frequently, this
joining technique transfigures the metallurgical structures, which therefore influences
the mechanical properties of the welded part. Consequently, crucial proficiency of
process-microstructure-properties is essential. **Table 2.3** shows examples of different
types of process.

In laser welding, the temperature-dependent yield strength is dependent on the
filler metal utilized. Meanwhile, other researchers considered the flow stresses of
martensite, ferrite, and austenite [38–40]. Liu et al. proposed a Voce-Ludwik fabrication model of strain hardening and a Johnson-Cook model for temperature sensitivity
[41]. The authors also studied the strain rate on the hardening behaviour of DP600
dual phase steel, and the results demonstrated that the strain rate had little or no effect
on residual stress in laser welding of DP600 steel [41].

It is important to include the numerical models and computational implementations
in the laser welding method as it is related to the metal behaviours of strain hardening
and temperature sensitivity. The rate-independent thermo-elastic finite element analysis of laser welding can be defined by the total strain increment [42] as described
in Equations 1–4:

$$d\varepsilon = d\varepsilon_e + d\varepsilon_T + d\varepsilon_p \tag{1}$$

**TABLE 2.3**
**Some examples of different types of laser welding processes**

| Process | Laser type | Joint materials composition | Laser power-Watt (W) | Substrate thickness (mm) | Conclusion | Reference |
|---|---|---|---|---|---|---|
| Pulse wave laser process | Nd:YAG | 304 stainless steel plates | ~1100 | 3 | Average power had insignificant effect on decreasing time | [48] |
| High-power laser welding | Fibre laser (IPG YLS) | 304 stainless steel | 10k | 15 | 3D model and adaptive mesh refinement method were formulated | [49] |
| High-power diode laser (HPDL) | Rofin-Sinar DL015 | AISI 1040 | 1500 | 6 | The HAZ depended on the amount of fluence distributed to the substrate during laser treatment | [50] |
| Laser welding | Not provided | Ultra-high strength steel 1700MS from SSAB of Sweden | 4000 | 2 | Temperature sensitivity and annealing increased the chances for plastic deformation, while strain hardening declined | [42] |
| Laser welding using high-power disc laser at green wavelength | TruDisk 1020 of TRUMPF Laser GmbH | Copper materials | 1k | 1 | It was discovered that copper vapour should be kept out of the beam path, to allow deep penetration welding process | [51] |
| Laser welding with lateral beam oscillation | TruDisk 12002D (Optical fibre-Ø 200μm) | Al EN AW-1050A and mild steel DC01 base materials | 2–4k | 1.5 | Beam oscillation increased the process tolerance influencing weld depth and seam shape | [52] |
| Twin-spot Laser welding | TruDisk 12002 laser, i.e. Yb:YAG solid-state laser (Trumpf) | Transformation Induced Plasticity Steel | 2–6k | 2 | Hardness of the fusion zone amounted to the hardness of the base material when using energy of 0.150kJ/mm. This value was due to the coarse-lath microstructure of the zone | [53] |

*(Continued)*

**TABLE 2.3 (Continued)**
**Some examples of different types of laser welding processes**

| Process | Laser type | Joint materials composition | Laser power- Watt (W) | Substrate thickness (mm) | Conclusion | Reference |
|---|---|---|---|---|---|---|
| Continuous-Wave Laser welding | IPG YLR-1000-SM single-mode fibre laser | Al-Cu alloy | 1k | 0.3–0.4 | When using 1–2-bead configuration, joint strength increased by 20%, while oscillating from 2–3-bead configuration led to negligible increases | [54] |
| Conventional Laser welding | High-power diode laser (HPDL) | Dissimilar NiTi wires | 1500 | 1.5 | Coarse grain structures were observed on the melting zone more than other sections exhibiting thermal effect and poor thermal conductivity of the NiTi alloys | [55] |
| Conventional Laser welding | Pulsed Nd:YAG laser | High strength steel of Inconel alloy 718 and AISI4140 steels | 1–2k | 2 | When welding parameters are increased, the output values are reduced exhibiting good mechanical properties of the welds | [56] |
| Dual-Laser beam welding | $CO_2$ laser | Al-Li alloys | 1.4–2k | 3.5 | The developed coupled model provided a tool in guiding a process and filler selection, and factors that contribute to hot cracking susceptibility | [57] |
| Laser beam welding | Nd:YAG laser | Martensitic (SUS420J2) and Austenitic (SUS303) stainless steels | 400 | – | Welding joint possess good mechanical performance under low temperature. When the temperature is reduced from 25–60 °C, the fracture force of weld joint increased from 7.5 to 9kN | [58] |
| Autogenous laser welding | Disk laser TRUMPF TruDisk 3302 (Yb:YAG) | STRENX 1100MC steel grade | 3.3 | 5 | Indifferent to rapid cooling, no tendency of cracking or HAZ experienced on the weld metal | [59] |

$$d\varepsilon_e = D_e^{-1}d\sigma + SdT \tag{2}$$

$$d\varepsilon_T = \alpha dT \tag{3}$$

$$d\varepsilon_p = \lambda\left\{\frac{\partial f}{\partial\sigma}\right\} \tag{4}$$

where the elastic stiffness matrix De, yield function f and the matrix S, linear expansion coefficient α, flow stress σ(εp,T).

Magalhães et al. [43] proposed the consummate approach to employ volumetric heat flux distribution in the laser welding process, which is compared to the classical model recommended by Goldak and Akhlaghi [44]. The authors showed two ways of representing 2D heat distribution, viz. linear or Gaussian. Accordingly, the linear heat flux distribution over a circular domain is:

$$q''(x,y,t) = \frac{Q(t)}{\pi R^2}\left((x-ut)^2 + y^2\right) \tag{3}$$

where $q''$ is the heat flux, Q is the gross heat rate, R is the radius and u is the welding speed. Meanwhile, the Gaussian heat distribution is described by:

$$q(x,y,t) = \frac{Q(t)}{\pi R^2}e^{-3\frac{x-ut^2}{r^2}}e^{-3\frac{y^2}{r^2}} \tag{4}$$

The 3D heat distribution applied in this study was based on a conical distribution of power density as suggested in by Goldak and Akhlaghi [44]. The overall recommended methodology is a complementary tool to estimate the weld bead profile and thermal efficiency in low penetration laser welding [43].

On one hand, temperature field in the vicinity of laser welding pool at various welding parameters can be calculated by employing the macroscopic heat-transfer model. However, the following are prerequisites for modelling of the thermodynamic process during laser welding [45]:

- Melting and solidification process during welding are considered instead of the effect of vapour on the liquid metal flow which is neglected.
- The liquid metal flow in the laser welding pool is governed by heat buoyancy and Marangoni shear stress.
- The liquid phase in the laser welding pool is ideally Newtonian, laminar and incompressible.
- It is theorized that the keyhole in the welding pool is stable, while the pressure recoil on the keyhole wall is balanced by tension on the surface.

Based on the listed assumptions above, a 3D macroscopic model is proposed to calculate the heat transfer throughout the laser welding process. The prevalent equations (both mass and energy conservation) are as follows:

*Mass conservation*

$$\frac{\partial \rho}{\partial \rho} + \nabla.(\rho \vec{u}) = 0 \qquad (5)$$

where $\rho$ is the mass density and $u = (u, v, w)$ is the velocity field in the welding pool.

*Momentum conservation*

$$\frac{\partial}{\partial t}(\rho \vec{u}) + \rho.(\vec{u}.\nabla)\vec{u} = -\nabla p + \eta \nabla.\left[\left(\nabla \vec{u} + \nabla \vec{u}^T\right) - \frac{2}{3}\nabla.\vec{u}I\right] + \vec{S}_m \qquad (6)$$

where $p$ is the static pressure, $\eta$ is the dynamic viscosity, $S_m$ is the source terms and is defined as:

$$\vec{S}_M = \rho \vec{g} - \rho \vec{g}\beta(T - T_L) - c_1\frac{(1-f_1)^2}{f_1^3 + c_2}(\vec{u} - \vec{u}_0) \qquad (7)$$

where $\beta$ is the thermal expansion coefficient, $g$ is the gravitational acceleration, $T_L$ is the liquidus temperature $f_l$ is the liquid fraction, $c_l$ is a constant that defined the resistance when liquid metal flows through the mushy zone, while $c_2$ is a smaller constant to continue the calculation when $f_l$ is close to zero [46].

*Energy conservation*

$$\frac{\partial}{\partial t}(\rho h) + \nabla.(\rho h \vec{u}) = \nabla.\left(\frac{K_s}{C_p}\nabla h\right) + S_h \qquad (8)$$

Where $K_s$ is the thermal conductivity, $h$ is the enthalpy and $c_p$ is the specific heat. Moreover, $S_h$ is the source term and is given by:

$$S_h = -\frac{\partial(\rho \Delta h)}{\partial t} - \nabla(\rho \vec{u} \Delta h) \qquad (9)$$

In a similar fashion, dynamic characteristics of the keyhole and weld pool in the underwater laser process was presented by Luo et al. [47] using a novel mathematical model. The findings indicated that when the water pressure is elevated, the keyhole surface temperature is also increased. As a result, a more stable and shallower keyhole is generated in the underwater laser welding process under 5MPa pressure.

## 2.4   LASER ANNEALING AND HARDENING

Laser melting (LM) is a computer-assisted device based additive manufacturing (AM) technique that employs a layer-by-layer building process to merge powder particles using local melt solidification by scanning laser beam of high energy density

[60–62]. This is a prospective process for the fabrication of metallic parts of complex shape geometries, in a short time span with no necessity of post machining. The local heating and fast cooling induced in the LM process result in the formation of sub-grains in the microstructure upon cooling. The work done by Saeidi et al. [4] found cellular structure formation with fine sub-grains with the size ~0.5μm during the LM process which were significantly smaller than the grain size in conventionally solidified 316L stainless steel. Further, the sub-grain boundaries exhibited a more substantial concentration of dislocation loops, while, their cells were enriched with Mo. The authors mentioned that the segregation of Mo in the cell boundaries and the resultant phases in the microstructure in the laser melted 316L stainless steel indicate non-equilibrium thermodynamic condition [4].

Normalizing, annealing, tempering, and hardening are frequently adapted to adjust microstructure and enhance mechanical properties [63]. Hardening and tempering strengthen the material by martensitic transformation, whereby the austenite transforms into hard/brittle martensite. The driving force for this hardening process is the presence of carbon in steel that accelerates the configuration of martensite. Martensite transformation during annealing is attributed to a favourable combination of hardness and strength in a variety of steel grades [64–70].

Laser surface hardening of AISI 420 stainless steel treated by a pulsed Nd:YAG laser was investigated [71]. The results indicated that laser treatment induces adsorption of the continuous carbide from grain boundaries into the grains, to make them spherical and discontinuous inside the grain and grain boundaries. This causes chromium carbides to dissolve, so a higher corrosion resistance can be achieved in laser-hardened area [71].

Selective laser melting (SLM) of 316L stainless steel displays columnar grains and the micro/nano size sub-grain or intercellular inside larger grains [72–74]. The deformation of this alloy is by twin-induced-plasticity (TWIP) [75] with amplified yield strength, compared with conventional 316L alloys [76].The strain rate dependent compressive flow stress of a SLM-316L alloy and C-316L alloy of the same grain sizes were investigated for comparison. The microstructural examinations of the cross-sections of deformed specimens indicated a TWIP in SLM-316L, while the C-316L showed martensitic transformations-induced-plasticity at both high strain and quasi-static rates [77]. The activation driving force in both alloys corresponds to dislocation intersections which showed a thermally activated deformation process (slip by dislocation) [77]. The effect of annealing temperature and strain rate on mechanical properties of a SLMed 316L alloy has been studied by Jiang et al. [78]. The authors found that the SLM-produced components annealed at 873 K led to mechanical performance compared to the as-built component, whereas thermal heat-treating at 1123K, contributed to the classical strength-ductility trade-off feature [78].

A repetitive laser scanning method to produce a 3D functional graded NiTi alloy by the SLM technique has been studied by Yang et al. [79]. The functionality of the graded alloy was analyzed by mechanical recoverable strain and deformation mechanisms. The gradient area showed B19 'phase deforming via the re-orientation of martensitic B19' variants in the early stages of deformation. The other gradient area comprised B2 phase which deformed via stress-induced martensitic modification at large applied stress [79]. These overlay multi-deformation mechanisms exhibit persistent increase of recoverable strain and exceptional strain hardening effect [79].

Mechanical and case-hardening properties of 16MnCr5 produced by LPBF has been investigated by characterizing the effect of the baseplate heating via shielding gas [80]. The results indicated that the higher the baseplate temperature, the component distortion declines substantially; however, the as-built material properties decrease. The microstructure showed grain coarsening due to tempering effects and recrystallization, which was attributed to loss of hardness and ultimate tensile strength by 50%. The case-hardening properties were influenced by baseplate of 400 °C and above, while nitrogen as a shielding gas alters the thermal balance, since thermal conductivity of nitrogen induces higher cooling rates [80].

High entropy alloys (HEAs) and medium entropy alloys (MEAs) are materials comprised of multiple principal elements with an equiatomic or near-equiatomic ratio between 5 at.% and 35 at.% [81, 82]. The alloys display desirable mechanical properties, viz. high strength and hardness, good corrosion resistance, and excellent high-temperature properties [83–86]. Han et al. attempted to indicate the possibility of obtaining the expected strength-ductility synergy of additively manufactured HEAs/MEAs by modifying microstructural heterogeneity using collateral heat-treatment [87].

## 2.5   SURFACE TREATMENT, COATING USING LASERS

Developing technologies in the field of green energy is growing around the world. It has been found that electrically conductive materials have been implemented for this purpose of green energetics performing good mechanical and chemical properties [27]. It has been demonstrated that the NiAl intermetallics are a strong material component of the goal of green energy. These alloy materials have been used as protective coatings for steel electrodes in direct current (DC) electrolysis methods for hydrogen production [88]. They are also utilized for molten carbon fuel cells (MCFC) operating at intermediate temperatures ~600 °C. Due to excellent heat-resistant properties of the NiAl alloy, it is favourable for use as a metallic current collector incorporated into the anodic structure in the MCFC. Another green energy field of NiAl materials is in the water-activated batteries, a technology that does not include electrolytes but generates voltage once fully submerged in water for a period of time [88].

The base of the functional graded and intermetallic coatings can be produced by a variety of thermal additive approaches, viz. pulse plasma sintering [89], reactive hot compaction of Ni and Al powders [90], plasma spray [91], and layerwise selective laser sintering [92, 93]. These processes are often characterized by incomplete solid-state reaction (heterogeneity) due to synthesis of intermetallic compounds like $Al_3Ni_2$, AlNi, $Al_3Ni$, also during high-speed crystallization of NiAl alloy materials, i.e. the semi-melting of NiAl particles occurring in the background degenerate the perspective of their future industrial applications [94].

Cold spraying (CS) is a widely used non-thermal spray coating process that is known for its free-form fabrication including all-solid state that uses high kinetic energy. The measurement of particle velocity prior to contact on the substrate is a key component of the CS process [95]. Anatolyevich and Vladimirovich [96] studied the laser post annealing of CSed NiAl composites coatings, and they found that the utilized specific heat was inadequate for remelting of all materials at 40W and 10mm/s laser spot velocity parameters. Further, due to thermal stresses built, the resultant cracks and pores inhibited heat spreading inside the coating allowing the formation of the eutectic NiAl dendritic boundary. However, the authors designed a system that exhibits the formation of a range of complex-shaped parts, with the exception of deep-holes and inner surfaces [96].

## 2.6   LASER MICRO AND NANO FABRICATION

Laser-based fabrication techniques have been used to construct medical devices with microscale and sub-microscale characteristics. The ability to produce small-length scales using short-production time-leads imparts the laser direct writing method with exclusive proficiency for the fabrication of medical devices. The technology has successfully processed a wide variety of advanced medical devices including patient-specific prostheses, biosensors, stents, drug-delivery devices, and tissue-engineering scaffolds [97]. Selective laser sintering has been used in synergistic manufacture of micro-nano bioactive tailored ceramic in bone tissue engineering. Xu et al. studied the micro/nano surface structure of polycaprolactone polylactic acid-nano hydroxyapatite (PCL-PLA-nHA) ternary porous scaffolds fabricated by selective laser sintering and in-situ hydrothermal deposition processes. The overall study indicated that the scaffold presented superior cytocompatibilty properties [98].

Laser shock-peening (LSP) has been used extensively as the surface treatment method to improve mechanical properties of engineering parts such as compressor discs and blades of jet engines [99–101]. Microstructural alterations near the surface are induced by the LSP, whereby a variety of deformed structures change the properties of treated materials. This generates the heterogeneous proportion of mechanical twins, dislocation structures, and grain refinement, including sub-grain and nanograin refinement [102]. Laser surface hardening of martensitic stainless steel was demonstrated for industrial application. Tani et al., [103] demonstrated that process optimization by conducting optimal process parameters in terms of laser parameters and scanning surface strategy on a AISI 420. The results showed that experimental activity can be reduced confirmed by the numerical estimated results. Komanduri and Hou [104] mentioned that analytical solutions exhibit precise estimations and render insight into the physical process of laser-transformation hardening of steels. Hardening of laser processed Al-8%Ca eutectic Al alloy showed increase in hardness due to formation of (1) supersaturated solid solution of calcium in Al and (2) dislocation loops under non-equilibrium conditions [105]. The influence of laser hardening of previously heat-treated 4Kh5MFS steel on microstructure, surface microhardness, and treated surface quality has been studied by Aborkin et al. [106]. The laser radiation power from 650 to 750W and the rate of treatment from 8 to 10 mm/sec have a significant influence on laser-hardening depth, microhardness, and surface quality, varying them in ranges of 0.64–0.86mm, 675–750HV, and 0.6–1.2μm, respectively [106].

Laser-assisted nano-imprinting lithography (LAN) can be conducted on a range of substrates; however, Chun-Ping Jen [107] proposed and demonstrated the feasibility of a metallic (copper) substrate instead of a conventionally used silicon. The reported results disclosed that the maximum temperature decreased at the surface of the copper substrate due to the increase in pulse duration. In comparison to the conventional silicon, the copper substrate absorbed higher energy from the excimer laser pulse (~20% more) owing to its larger thermal conductivity [107].

Silicon micro-nano structured electron field emitter arrays were manufactured employing pulsed krypton fluoride (KrF) excimer laser crystallization on a metal coated backplane sample outlined in the work by Mohammed Shamim et al. [108]. The technology delineated in this study signified its potential for prospective flat plane displays.

The prevailing optical technology is widely used in contemporary state of the art technology industries, such as, (1) communication factories with use of optical fibre and microwave communication; (2) computer industries with optical mass data-storage; (3) semiconductor industries such as photo-lithographic machines; and (4) mechanical engineering, e.g., laser machinery and clinical medical discipline with laser surgery [109]. For example, femtosecond laser processing is a universal method used for the fabrication of fibre optic sensors with its high processing precision rate and minuscule heat-affected zone developed in the structure. Innumerable optical fibre configurations have been established as gas-pressure sensors, for example, a diaphragm-ferrule-based sensor and microbubble-based Fabry-Perot Interferometer (FPI) used for high gas-pressure and temperature sensitivity [110] in conjunction with the hollow-core-based FPI [111], solid-core-based FPI [112] and fibre-tip micro-cavity pressure sensor [113] with good features, but there are challenges like cross-sensitivity to temperature to consider. Undoubtedly, wet etching-assisted femtosecond laser ablation adds to the economic value and great potential for the manufacturing of Ga-As in applications such as micro-supercomputer, microwave sensing, and optical communication industries [114]. The design and fabrication of novel honeycomb micro-texture should be taken into account for it is an imminent prospective technology that provides various micro-pool reservoirs for lubricants. The study outputs in [115] validated that the achieved wall-thickness improved the quantity of fins and their lengths which maximized heat-dissipation rate during machining.

## 2.7  CONCLUSIONS

In recent years, technological processes using diffractive free-form optics have been developed for applications in the aerospace industry. Research has shown that laser processing using diffractive free-form optics results in reduced porosity in coatings, and that there are excellent emerging ways for local annealing of coatings [116]. It is of the first magnitude to protect and strengthen the surface and repair the damaged part components in order to enhance their service life, diminishing material depletion thus promoting production efficiency [117].

It was shown that green laser radiation is the best method for welding materials that suffer unfavourable process proficiency as a result of supreme thermal conductivity, and low absorption coefficient for infrared wavelength, such as copper materials. The high-energy absorption capacity of low carbon medium manganese (LCMMn) steels fabricated by LPBF allows them to be used in the manufacture of energy-absorbing components such as the crash-box prototype profile. These components enhance crashworthiness by their materials geometry optimization [118].

The microstructure and corrosion effects of laser surface-melted high-speed steels has been studied extensively. The improvement of corrosion resistance in these steels is applicable to the combined influence of dissociation and refinement of large carbides and the increase of passivating alloying elements such as C, Mo, and W in solid solution [119].

Annealing and normalizing treatment in medium carbon steels do not affect or change microstructural phases, but rather changes in grain size. Laser surface treatment and/or tempering resulted in partial surface phase transformations [120]. A laser annealing optimization process of C45 steel using a finite element method (FEM) has been developed by Martinovs et al. [121]. It has been shown using numerical calculations that the number of annealing experiments can be reduced. The method can be used for other steel grades for calculating the change in temperature and other physical properties of the annealed surface during laser treatment. In essence, the established methodology can be utilized for hardening of parts that easily wear out, such as agricultural machines (ploughshares ploughing heavy clay soils) [121]. Softening of low-alloyed Ti billets with laser annealing was studied by Murzin and Kazanskiy [122]. Since it permits decrease of materials time during oxidation of surface layers at high gas saturation temperatures, laser heating/annealing was chosen for this work. The results also showed that laser annealing improved the ultimate tensile strain by 10–15% and decreased the bend angle by 30–40% for cold deformation of Ti-4Al-1.5Mn and Ti-2Al-1.5Mn alloys.

The precipitation hardening of selective laser melted (SLMed) AlSi10Mg alloy was examined in terms of solution treatments and ageing duration [123]. It was shown that the selection of solution heat-treatment (SHT) times determines the grain refinement of the alloy. For example, cast alloys require shorter SHT times to homogenize the finer microstructures. However, the fine structure of SLMed alloy parts require longer SHT times to stabilize the microstructure and improve mechanical properties with or without ageing [123]. Mechanical properties and microstructural evolution of cold-rolled FeCoCrNiMn-(N, Si) HEA fabricated by SLM was investigated by Zhang et al. [124]. It has been shown that doping with Ni and Si elements affected the microstructure at different annealing temperatures. The dispersion and small size of Cr2N particles enhanced the mechanical properties of the alloy by restricting the motion of dislocations and grain boundaries [124].

For the development of hybrid-alloy steel parts, additive depositioning of 18Ni300 steel powder to wrought 17-4 PH steel using the LPBF process was investigated by Chan et al. [125]. The results showed that this hybrid-build technique induces strong powder-to-substrate bonding with tensile ductility failure away from the interface on the side of the material with lower strength. This was indicative of the strong significance of solution strengthening within the fusion bond. It was also shown that this process is cost-effective, and an alternative for mould manufacturing industry-related processing [125].

## REFERENCES

[1]   N. S. Moghaddam, A. Jahadakbar, A. Amerinatanzi, & M. Elahinia, *Recent advances in laser-based additive manufacturing*, Taylor & Francis Group, 2017.

[2]   M. Q. Zafar & H. Y. Zhao, 4D Printing: Future insight in additive manufacturing, *Met. Mater. Int.*, 26, (5), 564–585, 2020.

[3]   J. G. Kim, J. B. Seol, J. M. Park, H. Sung, S. H. Park, & H. S. Kim, Effects of cell network structure on the strength of additively manufactured stainless steels, *Met. Mater. Int.*, 27, 2614–2622, 2021.

[4]   K. Saeidi, X. Gao, F. Lofaj, L. Kvetková, & Z. J. Shen, Transformation of austenite to duplex austenite-ferrite assembly in annealed stainless steel 316L consolidated by laser melting, *J. Alloys Compd.*, 633, 463–469, 2015, doi: 10.1016/j.jallcom.2015.01.249.

[5]   V. D. Divya, R. Munoz-Moreno, O. M. D. M. Messe, J. S. Barnard, S. Baker, T. Illston, & H. J. Stone, Microstructure of selective laser melted CM247LC nickel-based superalloy and its evolution through heat treatment, *Mater. Charact.*, 114, 62–74, 2016.

[6]   G. M. Karthik, E. S. Kim, A. Zargaran, P. Sathiyamoorthi, S. G. Jeong, & H. S. Kim, Role of cellular structure on deformation twinning and hetero-deformation induced strengthening of laser powder-bed fusion processed CuSn alloy, *Addit. Manuf.*, 54, (November 2021), 102744, 2022, doi: 10.1016/j.addma.2022.102744.

[7]   W. Steen & J. Mazumder, *Laser Material Processing*, 4th Edition. London: Springer, 2010.

[8]   J. Ion, *Laser Processing of Engineering Materials: Principles, Procedure and Industrial Application.* Jordan Hill, United Kingdom: Elsevier Science & Technology, 2005. [Online]. Available: http://ebookcentral.proquest.com/lib/pretoria-ebooks/detail.action?docID=294622

[9]   G. Chryssolouris, *Laser Machining: Theory and Practice.* New York: Springer Science & Business Media, 1991.

[10]  S. Darvekar, A. B. K. Rao, & S. S. Ganesh, Machining capability of a 2-D of parallel kinematic machine tool and conventional CNC milling machine, *Mater. Today Proc.*, 45, 3213–3218, 2021, doi: 10.1016/j.matpr.2020.12.376.

[11]  N. K. Angwenyi, N. M. Senga, N. K. Ronoh, F. M. Mwema, E. T. Akinlabi, & B. Tanya, The effects of machining parameters on conventional machining: An overview, *Mater. Today Proc.*, 44, 1540–1542, 2021, doi: 10.1016/j.matpr.2020.11.751.

[12]  F. Amirkhani, A. Dashti, H. Abedsoltan, A. H. Mohammadi, A. G. Chofreh, F. A. Goni, & J. J. Klemeš, Estimating flashpoints of fuels and chemical compounds using hybrid machine-learning techniques, *Fuel*, 323, (April), 2022, doi: 10.1016/j.fuel.2022.124292.

[13]  L. L. Thornton, D. E. Carlson, & M. R. Wiesner, Predicting emerging chemical content in consumer products using machine learning, *Sci. Total Environ.*, 834, (April), 154849, 2022, doi: 10.1016/j.scitotenv.2022.154849.

[14]  W. Cao, D. Wang, G. Cui, J. Zhang, & D. Zhu, Improvement on the machining accuracy of titanium alloy casing during counter-rotating electrochemical machining by using an insulation coating, *Surf. Coatings Technol.*, 443, 128585, 2022, doi: 10.1016/j.surfcoat.2022.128585.

[15]  J. Xue, B. Dong, & Y. Zhao, Significance of waveform design to achieve bipolar electrochemical jet machining of passivating material via regulation of electrode reaction kinetics, *Int. J. Mach. Tools Manuf.*, 177, (1088), 103886, 2022, doi: 10.1016/j.ijmachtools.2022.103886.

[16] M. J. Islam, Y. Zhang, L. Zhao, W. Yang, & H. Bian, Material wear of the tool electrode and metal workpiece in electrochemical discharge machining, *Wear,* 500–501 (April), 204346, 2022, doi: 10.1016/j.wear.2022.204346.

[17] M. Kowalczyk & K. Tomczyk, Assessment of measurement uncertainties for energy signals stimulating the selected NiTi alloys during the wire electrical discharge machining, *Precis. Eng.,* 76, (February), 133–140, 2022, doi: 10.1016/j.precisioneng.2022.03.005.

[18] A. Klink, S. Schneider, & T. Bergs, Development of a process signature for electrical discharge machining, *CIRP Ann.,* 00, 2–5, 2022, doi: 10.1016/j.cirp.2022.03.043.

[19] A. Z. Juri, Y. Zhang, A. Kotousov, & L. Yin, Zirconia responses to edge chipping damage induced in conventional and ultrasonic vibration-assisted diamond machining, *J. Mater. Res. Technol.,* 13, 573–589, 2021, doi: 10.1016/j.jmrt.2021.05.005.

[20] S. G. Kim, C. M. Lee, & D. H. Kim, Plasma-assisted machining characteristics of wire arc additive manufactured stainless steel with different deposition directions, *J. Mater. Res. Technol.,* 15, 3016–3027, 2021, doi: 10.1016/j.jmrt.2021.09.130.

[21] H. Pothur, V. Reddy, & R. Ganesan, Experimental investigations on process parameters of stainless steel 410 alloy by plasma arc machining process using grey relational analysis with entropy measurement, *Mater. Today Proc., (xxxx),* 2022, doi: 10.1016/j.matpr.2022.03.592.

[22] S. R. Mangaraj, D. K. Bagal, N. Parhi, S. N. Panda, A. Barua, & S. Jeet, Experimental study of a portable plasma arc cutting system using hybrid RSM-nature inspired optimization technique, *Mater. Today Proc.,* 50, 867–878, 2021, doi: 10.1016/j.matpr.2021.06.138.

[23] V. B. Magdum, J. K. Kittur, & S. C. Kulkarni, Surface roughness optimization in laser machining of stainless steel 304 using response surface methodology, *Mater. Today Proc.,* 59, 540–546, 2022, doi: 10.1016/j.matpr.2021.11.570.

[24] K. E. Hazzan & M. Pacella, A novel laser machining strategy for cutting tool repair, *Manuf. Lett.,* 32, 87–91, 2022, doi: 10.1016/j.mfglet.2022.04.005.

[25] S. Santosh, J. Kevin Thomas, M. Pavithran, G. Nithyanandh, & J. Ashwath, An experimental analysis on the influence of $CO_2$ laser machining parameters on a copper-based shape memory alloy, *Opt. Laser Technol.,* 153, (April), 108210, 2022, doi: 10.1016/j.optlastec.2022.108210.

[26] A. N. Samant & N. B. Dahotre, Laser machining of structural ceramics a review, *J. Eur. Ceram. Soc,* 29, (6), 969–993, 2009.

[27] N. Dahotre & A. Samant, *Laser machining of advanced materials.* Ist edition, CRC Press, 236, 2011, eBook ISBN 9780429106736, doi: 10.1201/b10862.

[28] H. D. Vora, S. Santhanakrishnan, S. P. Harimkar, S. K. S. Boetcher, & N. B. Dahotre, Evolution of surface topography in one-dimensional laser machining of structural alumina, *J. Eur. Ceram. Soc.,* 32, (16), 4205–4218, 2012, doi: 10.1016/j.jeurceramsoc.2012.06.015.

[29] H. D. Vora, S. Santhanakrishnan, S. P. Harimkar, S. K. S. Boetcher, & N. B. Dahotre, One-dimensional multipulse laser machining of structural alumina: evolution of surface topography, *International Journal of Advanced Manufacturing Technology,* 68, (16), 69–83, 2012, doi: 10.1016/j.jeurceramsoc.2012.06.015.

[30] J. Yang, J. Yu, Y. Cui, & Y. Huang, New laser machining technology of Al 2O 3 ceramic with complex shape, *Ceram. Int.,* 38, (5), 3643–3648, 2012, doi: 10.1016/j.ceramint.2012.01.003.

[31] S. Marimuthu, J. Dunleavey, & B. Smith, Picosecond laser machining of tungsten carbide, *Int. J. Refract. Met. Hard Mater.,* 92, (July), 105338, 2020, doi: 10.1016/j.ijrmhm.2020.105338.

[32] H. Li, F. Wen, Y. S. Wong, F. Y. C. Boey, V. S. Subbu, D. T. Leong, K. W. Ng, G. K. L. Ng, & L. P. Tan, Direct laser machining-induced topographic pattern promotes up-regulation of myogenic markers in human mesenchymal stem cells, *Acta Biomater.*, 8, (2), 531–539, 2012, doi: 10.1016/j.actbio.2011.09.029.

[33] D. Sun, F. Han, W. Ying, & C. Jin, Surface integrity of water jet guided laser machining of CFRP, *Procedia CIRP*, 71, 71–74, 2018, doi: 10.1016/j.procir.2018.05.073.

[34] B. Guimarães, D. Figueiredo, C. M. Fernandes, F. S. Silva, G. Miranda, & O. Carvalho, Laser machining of WC-Co green compacts for cutting tools manufac-turing, *Int. J. Refract. Met. Hard Mater.,* 81, (December 2018), 316–324, 2019, doi: 10.1016/j.ijrmhm.2019.03.018.

[35] Y. Wu, A. Y. Vorobyev, R. L. Clark, & C. Guo, Femtosecond laser machining of electrospun membranes, *Appl. Surf. Sci.,* 257, (7), 2432–2435, 2011, doi: 10.1016/j.apsusc.2010.09.111.

[36] J. K. Park, J. W. Yoon, & S. H. Cho, Vibration assisted femtosecond laser machining on metal, *Opt. Lasers Eng.*, 50, (6), 833–837, 2012, doi: 10.1016/j.optlaseng.2012.01.017.

[37] W. Feng, J. Guo, W. Yan, Y. C. Wan, & H. Zheng, Deep channel fabrication on copper by multi-scan underwater laser machining, *Opt. Laser Technol.,* 111, 653–663, 2019, doi: 10.1016/j.optlastec.2018.10.046.

[38] C. Heinze, A. Pittner, M. Rethmeier, & S. S. Babu, Dependency of martensite start temperature on prior austenite grain size and its influence on welding-induced residual stresses, *Comput. Mater. Sci.*, 69, 60, 2013.

[39] J. Sun, J. Hensel, J. Klassen, T. Nitschke-Pagel, & K. Dilger, Solidstate phase trans-formation and strain hardening on the residual stresses in S355 steel weldments., *J. Mater. Process. Technol.*, 265, 84, 2019.

[40] S. A. Tsirkas, P. Papanikos, & T. Kermanidis, Numerical simulation of the laser welding process in butt-joint specimens, *J. Mater. Process. Technol.,* 134, 59–69, 2003.

[41] S. Liu, A. Kouadri-Henni, & A. Gavrus, Numerical simulation and experimental investigation on the residual stresses in a laser beam welded dual phase DP600 steel plate: thermomechanical material plasticity model, *Int. J. Mech. Sci.*, 122, 235–343, 2017.

[42] J. Xu, L. Wang, R. Wu, Y. Huang, & Y. Rong, Stress evolution mechanism during laser welding of ultra-high-strength steel: considering the effects of temperature sen-sitivity, strain hardening and annealing, *J. Mater. Res. Technol.*, 19, 1711–1723, 2022, doi: 10.1016/j.jmrt.2022.05.160.

[43] E. dos S. Magalhães, L. E. dos S. Paes, M. Pereira, C. A. da Silveira, A. de S. P. Pereira, & S. M. M. Lima e Silva, A thermal analysis in laser welding using inverse problems, *Int. Commun. Heat Mass Transf.,* 92, (March), 112–119, 2018, doi: 10.1016/j.icheat masstransfer.2018.02.014.

[44] J. A. Goldak & M. Akhlaghi, Computational welding mechanics, *Comput. Weld. Mech.*, 1–321, 2005, doi: 10.1007/b101137.

[45] L. Wang, Y. Wei, J. Chen, & W. Zhao, Macro-micro modeling and simulation on col-umnar grains growth in the laser welding pool of aluminum alloy, *Int. J. Heat Mass Transf.*, 123, 826–838, 2018, doi: 10.1016/j.ijheatmasstransfer.2018.03.037.

[46] A. D. Brent, V. R. Voller, & K. J. Reid, Enthalpy-porosity technique for modeling convection-diffusion phase change: Application to the melting of a pure metal, *Numer. Heat Transf.* 13, (3), 297–318, 1988.

[47] M. Luo, R. Hu, Q. Li, A. Huang, & S. Pang, Physical understanding of keyhole and weld pool dynamics in laser welding under different water pressures, *Int. J. Heat Mass Transf.*, 137, 328–336, 2019, doi: 10.1016/j.ijheatmasstransfer.2019.03.129.

[48] T. Liu, L. Yang, S. Zhao, & Y. Huang, Electrical Characteristics of Plasma Plume During Pulse Wave Laser Welding, *Trans. Tianjin Univ.*, 25, (4), 420–428, 2019, doi: 10.1007/s12209-018-0176-0.

[49] T. Liu, R. Hu, X. Chen, S. Gong, & S. Pang, Localized boiling-induced spatters in the high-power laser welding of stainless steel: Three-dimensional visualization and physical understanding, *Appl. Phys. A Mater. Sci. Process.*, 124, (9), 1–14, 2018, doi: 10.1007/s00339-018-2058-7.

[50] S. Guarino, M. Barletta, & A. Afilal, High Power Diode Laser (HPDL) surface hardening of low carbon steel: Fatigue life improvement analysis, *J. Manuf. Process.*, 28, 266–271, 2017, doi: 10.1016/j.jmapro.2017.06.015.

[51] M. Haubold, A. Ganser, T. Eder, & M. F. Zäh, Laser welding of copper using a high power disc laser at green wavelength, *Procedia CIRP*, 74, 446–449, 2018, doi: 10.1016/j.procir.2018.08.161.

[52] C. Mittelstädt, T. Seefeld, P. Woizeschke, & F. Vollertsen, Laser welding of hidden T-joints with lateral beam oscillation, *Procedia CIRP*, 74, 456–460, 2018, doi: 10.1016/j.procir.2018.08.151.

[53] A. Grajcar, M. Morawiec, M. Różański, & S. Stano, Twin-spot laser welding of advanced high-strength multiphase microstructure steel, *Opt. Laser Technol.*, 92, (January), 52–61, 2017, doi: 10.1016/j.optlastec.2017.01.011.

[54] A. Fortunato & A. Ascari, Laser welding of thin copper and aluminum sheets: Feasibility and challenges in continuous-wave welding of dissimilar metals, *Lasers Manuf. Mater. Process.*, 6, (2), 136–157, 2019, doi: 10.1007/s40516-019-00085-z.

[55] M. Mehrpouya, A. Gisario, M. Barletta, S. Natali, & F. Veniali, Dissimilar Laser Welding of NiTi Wires, *Lasers Manuf. Mater. Process.*, 6, (2), 99–112, 2019, doi: 10.1007/s40516-019-00084-0.

[56] M. Anuradha, V. C. Das, D. Venkateswarlu, & M. Cheepu, Parameter optimization for laser welding of high strength dissimilar materials, *Mater. Sci. Forum*, 969, 558–564, 2019, doi: 10.4028/www.scientific.net/MSF.969.558.

[57] Y. Tian, J. D. Robson, S. Riekehr, N. Kashaev, L. Wang, T. Lowe, & A. Karanika, Process Optimization of Dual-Laser Beam Welding of Advanced Al-Li Alloys Through Hot Cracking Susceptibility Modeling, *Metall. Mater. Trans. A Phys. Metall. Mater. Sci.*, 47, (7), 3533–3544, 2016, doi: 10.1007/s11661-016-3509-4.

[58] W. W. Zhang & S. Cong, Process optimization and performance evaluation on laser beam welding of austenitic/martensitic dissimilar materials, *Int. J. Adv. Manuf. Technol.*, 92, (9–12), 4161–4168, 2017, doi: 10.1007/s00170-017-0513-9.

[59] A. Kurc-Lisiecka, J. Piwnik, & A. Lisiecki, Laser welding of new grade of advanced high strength steel STRENX 1100 MC, *Arch. Metall. Mater.*, 62, (3), 1651–1657, 2017, doi: 10.1515/amm-2017-0253.

[60] J. D. Majumdar, A. Pinkerton, Z. Liu, I. Manna, & L. Li, Microstructure characterisation and process optimization of laser assisted rapid fabrication of 316l stainless steel, *Appl. Surf. Sci*, 247, 320–327, 2005.

[61] G. N. Levy, The role and future of the laser technology in the additive manufacturing environment, *Phys. Procedia*, 5, 65–80, 2010.

[62] F. Feuerhahn, Manurfacture and properties of selective laser melted high hardness tool steel, *Phys. Procedia*, 41, 836–841, 2013.

[63] A. Raymond & B. Higgins, *Properties of Engineering Materials*, second edition. London: Butterworth-Heinemann, 1985.

[64] X. L. Xu & F. Liu, Crystal growth due to recrystallization upon annealing rapid solidification microstructures of deeply undercooled single phase alloys quenched before recalescence, *Cryst. Growth Des.*, 14, 2110–2114, 2014.

[65] B. Kumar & S. Sharma, Recrystallization characteristics of cold rolled austenitic stainless steel during repeated annealing, *Mater. Sci. Forum*, 753, 157–162, 2013.

[66] C. J. Múnez, M. V. Utrilla, & A. Ureña, Effect of temperature on sintered austeno–ferritic stainless steel microstructure, *J. Alloy. Comp.*, 463, 552–558, 2008.

[67] C. García & F. Martín, Effect of ageing heat treatments on the microstructure and intergranular corrosion of powder metallurgy duplex stainless steels, *Corros. Sci.*, 52, 3725–3737, 2010.

[68] M. Campos & J. M. Torralba, Surface assessment in low alloyed Cr–Mo sintered steels after heat and thermochemical treatment, *Surf. Coat. Technol.*, 182, 351–362, 2004.

[69] S. Z. Qamar, Effect of heat treatment on mechanical properties of H11 tool steel, *J. Achiev. Mater. Manuf. Eng.*, 25, 115–120, 2009.

[70] D. A. Fadare, T. G. Fadara, & O. Y. Akanbi, Effect of heat treatment on mechanical properties and microstructure of NST 37-2 steel, *J. Min. Mater. Charact. Eng.*, 10, 299–308, 2011.

[71] B. Mahmoudi, M. J. Torkamany, A. R. S. R. Aghdam, & J. Sabbaghzade, Laser surface hardening of AISI 420 stainless steel treated by pulsed Nd:YAG laser, *Mater. Des.*, 31, (5), 2553–2560, 2010, doi: 10.1016/j.matdes.2009.11.034.

[72] Y. J. Yin, J. Q. Sun, J. Guo, X. F. Kan, & D. C. Yang, Mechanism of high yield strength and yield ratio of 316 L stainless steel by additive manufacturing, *Mater. Sci. Eng. A-Struct. Mater. Prop. Microstruct. Process.*, 744, 773–777, 2019.

[73] X. L. Wang, J. A. Muniz-Lerma, M. A. Shandiz, O. Sanchez-Mata, & M. Brochu, Crystallographic-orientation-dependent tensile behaviours of stainless steel 316L fabricated by laser powder bed fusion, *Mater. Sci. Eng. A-Struct. Mater. Prop. Microstruct. Process.*, 766, 16, 2019.

[74] M. M. Ma, Z. M. Wang, & X. Y. Zeng, A comparison on metallurgical behaviors of 316L stainless steel by selective laser melting and laser cladding deposition, *Mater. Sci. Eng. A-Struct. Mater. Prop. Microstruct. Process.*, 685, 265–273, 2017.

[75] W. Woo, J. S. Jeong, D. K. Kim, C. M. Lee, S. H. Choi, J. Y. Suh, S. Y. Lee, S. Harjo, & T. Kawasaki, Stacking fault energy analyses of additively manufactured stainless steel 316L and CrCoNi medium entropy alloy using in situ neutron diffraction, *Sci. Rep.*, 10, (1), 15, 2020.

[76] A. Rottger, K. Geenen, M. Windmann, F. Binner, & W. Theisen, Comparison of microstructure and mechanical properties of 316 L austenitic steel processed by selective laser melting with hot-isostatic pressed and cast material, *Mater. Sci. Eng. A-Struct. Mater. Prop. Microstruct. Process.*, 678, 365–376, 2016.

[77] M. Güden, S. Enser, M. Bayhan, A. Taşdemirci, & H. Yavaş, The strain rate sensitive flow stresses and constitutive equations of a selective-laser-melt and an annealed-rolled 316L stainless steel: A comparative study, *Mater. Sci. Eng. A,* 838, (January), 2022, doi: 10.1016/j.msea.2022.142743.

[78] H. Z. Jiang, Z. Y. Li, T. Feng, P. Y. Wu, Q. S. Chen, S. K. Yao, & J. Y. Hou, Effect of Annealing Temperature and Strain Rate on Mechanical Property of a Selective Laser Melted 316L Stainless Steel, *Acta Metall. Sin. (English Lett.,* 35, (5), 773–789, 2022, doi: 10.1007/s40195-021-01342-x.

[79] Y. Yang, J. B. Zhan, J. B. Sui, C. Q. Li, K. Yang, P. Castany, & T. Gloriant, Functionally graded NiTi alloy with exceptional strain-hardening effect fabricated by SLM method, *Scr. Mater.*, 188, 130–134, 2020, doi: 10.1016/j.scriptamat.2020.07.019.

[80] M. Schmitt, B. Kempter, I. Syed, A. Gottwalt, M. Horn, M. Binder, J. Winkler, G. Schlick, T. Tobie, K. Stahl, & G. Reinhart, Influence of baseplate heating and shielding

gas on distortion, mechanical and case hardening properties of 16MnCr5 fabricated by laser powder bed fusion, *Procedia CIRP,* 93, 581–586, 2020, doi: 10.1016/j.procir.2020.03.089.

[81]   D. B. Miracle & O. N. Senkov, A critical review of high entropy alloys and related concepts, *Acta Mater.*, 122, 448–511, 2017.

[82]   J. W. Yeh, S. K Chen, S. J. Lin, J. Y. Gan, T. S. Chin, T. T. Shun, C. H. Tsau, & S. Y. Chang, Nanostructured high-entropy alloys with multiple principal elements: Novel alloy design concepts and outcomes, *Adv. Eng. Mater.,* 6, 299–303, 2004.

[83]   H. Luo, S. S. Sohn, W. Lu, L. Li, X. Li, C. K. Soundararajan, W. Krieger, Z. Li, & D. Raabe, A strong and ductile medium-entropy alloy resists hydrogen embrittlement and corrosion, *Nat. Commun.,* 11, 1–8, 2020.

[84]   Q. Ye, K. Feng, Z. Li, F. Lu, R. Li, J. Huang, & Y. Wu, Microstructure and corrosion properties of CrMnFeCoNi high entropy alloy coating, *Appl. Surf. Sci.*, 396, 1420–1426, 2017.

[85]   W. R. Wang, W. L. Wang, S. C. Wang, Y. C. Tsai, C. H. Lai, & J. W. Yeh, Effects of Al addition on the microstructure and mechanical property of Al xCoCrFeNi highentropy alloys, *Intermetallics*, 26, 44–51, 2012.

[86]   Y. Zhang, T. T. Zuo, Z. Tang, M. C. Gao, K. A. Dahmen, P. K. Liaw, & Z. P. Lu, Microstructures and properties of high-entropy alloys, *Prog. Mater. Sci.,* 61, 1–93, 2014.

[87]   T. Han, Y. Liu, D. Yang, N. Qu, M. Liao, Z. Lai, M. Jiang, & J. Zhu, Effect of annealing on microstructure and mechanical properties of AlCrFe2Ni2 medium entropy alloy fabricated by laser powder bed fusion additive manufacturing, *Mater. Sci. Eng. A,* 839, (November 2021), 142868, 2022, doi: 10.1016/j.msea.2022.142868.

[88]   P. Aizpurietis, M. Vanags, J. Kleperis, & G. Bajars, NiAl protective coating of steel electrodes in DC electrolysis for hydrogen production/Ni-Al, *Latv. J. Phys. Tech. Sci.*, 50, (3), 53–59, 2013.

[89]   H. Zhu & R. Abbaschian, Microstructures and properties of in-situ NiAl-Al2O3 functionally gradient composites, *Compos. Part B Eng.,* 31, (5), 383–390, 2000.

[90]   H. Lee, S. Jung, S. Lee, & K. Ko, Alloying of cold-spreyed Al-Ni composite coatings by post-annealing, *Appl. Surf. Sci,* 253, (7), 3496–3502, 2007.

[91]   A. J. Michalski, J. J. M. Rosinski, & D. Siemiaszko, NiAl-Al2O3 composite produced by pulse plasma sintering with the participation of the SHS reaction, *Intermetallics*, 14, (6), 603–606, 2006.

[92]   B. S. Sidhu & S. Prakash, Evaluation of the corrosion behavior of plasma-sprayed Ni3Al coatings on steel in oxidation and molten salt environments at 900C, *Surf. Coat. Technol.,* 166, (1), 89–100, 2003.

[93]   I. Shishkovskii, A. G. Makarenko, & A. Petrov, Conditions for SHS of intermetallic compounds with selective laser sintering of powdered compositions, *Combust. Explos. Shock Waves,* 35, (2), 166–170, 1999.

[94]   S. N. Grigoriev, *Advanced Machining Technologies: Traditions and Innovations*, (vol. 834). Pfaffikon: Trans Tech Publications Ltd, 2015. [Online]. Available: https://search.ebscohost.com/login.aspx?direct=true&db=nlebk&AN=1107277&site=ehost-live&scope=site

[95]   A. Sova, S. Grigoriev, A. Okunkova, & I. Smurov, Potential of cold gas dynamic spray as additive manufacturing technology, *Int. J. Adv. Manuf. Technol.*, 69, (9–12), 2269–2278, 2013.

[96]   P. P. Anatolyevich & S. I. Vladimirovich, Laser post annealing of cold-sprayed Al-Ni composite coatings for green energy tasks, *Materials Sci. Forum*, 834, 113–118, 2015.

[97] S. D. Gittard & R. J. Narayan, Laser direct writing of micro- and nano-scale medical devices, *Expert Rev. Med. Devices*, 7, (3), 343–356, 2010, doi: 10.1586/erd.10.14.

[98] Y. Xu, W. Ding, M. Chen, H. Du, & T. Qin, Synergistic fabrication of micro-nano bioactive ceramic-optimized polymer scaffolds for bone tissue engineering by in situ hydrothermal deposition and selective laser sintering, *J. Biomater. Sci. Polym. Ed.*, 33, (16), 2104–2123, 2022, doi: 10.1080/09205063.2022.2096526.

[99] G. Lu, H. Liu, C. Lin, Z. Zhang, P. Shukla, Y. Zhang, & J. Ya, Improving the fretting performance of aero-engine tenon joint materials using surface strengthening, *Mater. Sci. Technol.*, 35, 1781–1788, 2019.

[100] U. Trdan, M. Skarba, & J. Grum, Laser shock peening effect on the dislocation transitions and grain refinement of Al-Mg-Si alloy, *Mater. Charact.*, 97, 57–68, 2014.

[101] Y. W. Fang, Y. H. Li, W. F. He, & P. Y. Li, Effects of laser shock processing with different parameters and ways on residual stresses fields of a TC4 alloy blade, *Mater. Sci. Eng. A*, 559, 683–692, 2013.

[102] R. Li, Y. Wang, N. Xu, Z. Yan, S. Li, M. Zhang, J. Almer, Y. Ren, & Y. D. Wang, Unveiling the origins of work-hardening enhancement and mechanical instability in laser shock peened titanium, *Acta Mater.*, 229, 117810, 2022, doi: 10.1016/j.actamat.2022.117810.

[103] G. Tani, A. Fortunato, A. Ascari, & G. Campana, Laser surface hardening of martensitic stainless steel hollow parts, *CIRP Ann. – Manuf. Technol.*, 59, (1), 207–210, 2010, doi: 10.1016/j.cirp.2010.03.077.

[104] R. Komanduri & Z. B. Hou, Thermal analysis of the laser surface transformation hardening process, *Int. J. Heat Mass Transf.*, 44, (15), 2845–2862, 2001, doi: 10.1016/S0017-9310(00)00316-1.

[105] S. O. Rogachev, E. A. Naumova, M. A. Vasina, N. Y. Tabachkova, N. V. Andreev, & A. A. Komissarov, Anomalous hardening of Al-8%Ca eutectic alloy due to a non-equilibrium phase state transition under laser irradiation, *Mater. Lett.*, 317, (March), 132129, 2022, doi: 10.1016/j.matlet.2022.132129.

[106] A. V. Aborkin, V. E. Vaganov, A. N. Shlegel', & I. M. Bukarev, Effect of Laser Hardening on Die Steel Microhardness and Surface Quality, *Metallurgist*, 59, (7–8), 619–625, 2015, doi: 10.1007/s11015-015-0148-8.

[107] C. P. Jen, Heat transfer analysis of laser-assisted nano-imprinting on a metallic substrate – Technical communication, *Mach. Sci. Technol.*, 12, (4), 563–577, 2008, doi: 10.1080/10910340802519312.

[108] M. Z. M. Shamim, S. Persheyev, M. Zaidi, M. Usman, M. Shiblee, S. J. Ali, & M. R. Rahman, Micro-Nano Fabrication of Self-Aligned Silicon Electron Field Emitter Arrays Using Pulsed KrF Laser Irradiation, *Integr. Ferroelectr.*, 204, (1), 47–57, 2020, doi: 10.1080/10584587.2019.1674988.

[109] S. Kawata, K. Okamoto, & S. Shoji, Photon-induced micro/nano fabrication, manipulation, and imaging with unconventional photo-active systems, *Mol. Cryst. Liq. Cryst. Sci. Technol. Sect. A Mol. Cryst. Liq. Cryst.*, 314, 173–178, 1998, doi: 10.1080/10587259808042475.

[110] T. She, H. Lin, F. Liu, Y. Dai, & D. Liu, A high sensitivity gas pressure sensor based on femtosecond laser micro/nano processing, *Integr. Ferroelectr.*, 197, (1), 70–76, 2019, doi: 10.1080/10584587.2019.1592083.

[111] L. Jin, B. Guan, & H. Wei, Sensitivity characteristics of Fabry-Perot pressure sensors based on hollow-core microstructured fibers, *J. Light. Technol.*, 31, (15), 2526, 2013.

[112] C. Wu, H. Y. Fu, K. K. Qureshi, B.-O. Guan, & H. Y. Tam, High-pressure and high-temperature characteristics of a Fabry–Perot interferometer based on photonic crystal fiber, *Opt. Lett.*, 36, (3), 412, 2011, doi: 10.1364/ OL.36.000412.

[113] J. Ma, J. Ju, L. Jin, & W. Jin, A compact fiber-tip micro-cavity sensor for high-pressure measurement, *IEEE Photon. Technol. Lett.*, 23, (21), 1561, 2011, doi: 10.1109/LPT.2011.2164060.

[114] X. Sun, F. Zhou, X. Dong, F. Zhang, C. Liang, L. Duan, Y. Hu, & J. Duan, Fabrication of GaAs micro-optical components using wet etching assisted femtosecond laser ablation, *J. Mod. Opt.*, 67, (20), 1516–1523, 2020, doi: 10.1080/09500340.2020.1869849.

[115] R. Sharma, S. Pradhan, & R. N. Bathe, Design and fabrication of honeycomb micro-texture using femtosecond laser machine, *Mater. Manuf. Process.*, 36, (11), 1314–1322, 2021, doi: 10.1080/10426914.2021.1906898.

[116] S. P. Murzin, N. L. Kazanskiy, & C. Stiglbrunner, Analysis of the advantages of laser processing of aerospace materials using diffractive optics, *Metals (Basel)*, 11, (6), 2021, doi: 10.3390/met11060963.

[117] Y. Peng, W. Zhang, T. Li, M. Zhang, L. Wang, & S. Hu, Microstructures and wear-resistance of WC-reinforced high entropy alloy composite coatings by plasma cladding: effect of WC morphology, *Surf. Eng.*, 37, (5), 678–687, 2021, doi: 10.1080/02670844.2020.1812480.

[118] A. Pawlak, R. Dziedzic, M. Kasprowicz, W. Stopyra, B. Kuźnicka, E. Chlebus, B. Schob, C. Zopp, L. Kroll, R. Kordass, & J. Bohlen, Properties of Medium-Manganese Steel Processed by Laser Powder Bed Fusion the Effect of the As-Built and Intercritically Annealed Microstructure on Energy Absorption During Tensile and Impact Tests, *Materials Science & Engineering A*, 870, 144859, 2023, doi.org/10.1016/j.msea.2023.144859.

[119] C. T. Kwok, F. T. Cheng, & H. C. Man, Microstructure and corrosion behavior of laser surface-melted high-speed steels, *Surf. Coatings Technol.*, 202, (2), 336–348, 2007, doi: 10.1016/j.surfcoat.2007.05.085.

[120] A. F. Hamood & S. A. Hafeed, Influence of Annealing, Normalizing Hardening Followed By Tempering And Laser Treatments on Some of The Static and Dynamic Mechanical Properties of Medium Carbon Steel المتبوع بال مراجعة والمعاملة الليزرية على متوسط الكربون أثر التلدين والتطبيع و الإطفاء, *Eng. Tech. J.*, 28, (21), 6274–6287, 2010.

[121] A. Martinovs, S. Polukoshko, E. Zaicevs, & R. Revalds, Laser hardening process optimization using FEM, *Eng. Rural Dev.*, 19, (June), 1500–1508, 2020, doi: 10.22616/ERDev2020.19.TF372.

[122] S. P. Murzin & N. L. Kazanskiy, Softening of Low-alloyed Titanium Billets with Laser Annealing, *IOP Conf. Ser. Mater. Sci. Eng.*, 302, (1), 2018, doi: 10.1088/1757-899X/302/1/012070.

[123] N. T. Aboulkhair, C. Tuck, I. Ashcroft, I. Maskery, & N. M. Everitt, On the Precipitation Hardening of Selective Laser Melted AlSi10Mg, *Metall. Mater. Trans. A Phys. Metall. Mater. Sci.*, 46, (8), 3337–3341, 2015, doi: 10.1007/s11661-015-2980-7.

[124] Z. Zhang, Y. Wu, J. Gu, & M. Song, Microstructural Evolution of a Selective Laser Melted FeCoCrNiMn–(N,Si) High-Entropy Alloy Subject to Cold-Rolling and Subsequent Annealing, *Adv. Eng. Mater.*, 2200131, 1–8, 2022, doi: 10.1002/adem.202200131.

[125] Y. L. S. Chan, O. Diegel, & X. Xu, Bonding integrity of hybrid 18Ni300-17-4 PH steel using the laser powder bed fusion process for the fabrication of plastic injection mould inserts, *Int. J. Adv. Manuf. Technol.*, 120, (7–8), 4963–4976, 2022, doi: 10.1007/s00170-022-09004-7.

# 3 Ingot Metallurgy Routes
## Conventional and Non-Conventional Casting Routes

*A. S. Bolokang and M. N Mathabathe*

## 3.1 INTRODUCTION

There are conventional and non-conventional casting routes which include the investment casting technique. The casting type depends on the kind of product required, and the mechanical properties and surface finish of the product. Several steps or processes are followed to produce a final casting. Vacuum arc melting (VAC) is discussed in this chapter. To produce alloys such as Ti which has high affinity to interstitial elements (oxygen, nitrogen, hydrogen), VAC is crucial in order to obtain high-purity metal and fewer defects in the casting. Moreover, the casting process can be very costly, especially during product development. In this process, repeated trials to obtain the best product may be costly in terms of raw materials and multiple melting. As a result, numerical tools combined with experimental tests are critical. Modelling and simulation before actual casting save costs and yield the best results. This may be beneficial especially when biomedical parts such as intricate prostheses are developed and made.

## 3.2 INVESTMENT CASTING

The investment casting (IC) process is an old casting method and uses the lost wax technique, which was used to produce jewellery and statues. In modern design and production of castings, the IC process has the following basic steps, shown in **Fig. 3.1**:

### 3.2.1 CASTING SIMULATION AND MODELLING

Numerical tools are used to predict imperfections in a casting prior to actual metal pouring into a designed mould. These computational tools predict where defects such as porosity are likely to occur. As a result, re-design of gating systems and risers are optimized before the melting and casting process. After optimization, the pattern is developed followed by moulding. These modern tools are easily printed nowadays by using 3D printing machines through additive manufacturing.

DOI: 10.1201/9781003356714-3

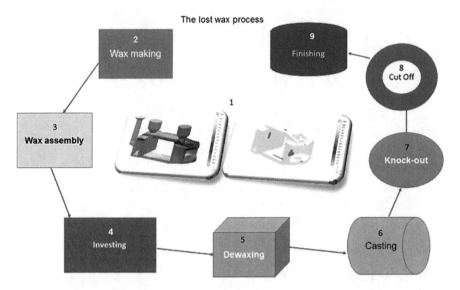

**FIG. 3.1**   Schematic illustration of the IC process

- Wax making – Wax patterns are produced by injection moulding. A die made of metal is used.
- Wax assembly – Through a gate, patterns are attached tree-like to a central wax sprue where a casting assembly is mounted on a pouring cup.
- Investing – Following wax construction, a shell is created by submerging the assembly in a liquid ceramic slurry, depositing numerous layers, and then submerging the assembly in a bed of ultrafine sand.
- Dewaxing – After the ceramic is dry, the wax is melted out in an autoclave, leaving a negative impression of the assembly inside the shell, which will be placed in an oven at high temperature. The shell represents a mould ready for the liquid metal.
- Casting – The mould/shell is filled with liquid to produce a solid casting. The casting is still attached to the ceramic shell.
- Knock out – After solidification and cooling of the casting, the ceramic shell is broken off by vibration or water blasting.
- Cut off – The parts are cut from the tree using approved cutting tools. Normally, the cutting will leave protrusions and uneven layers or edges on metals.
- Finishing – These are the cosmetic finishing operations like fettling and grinding. Care is taken not to induce surface defects or tamper with the dimensional accuracy of the castings.

**Fig. 3.2** indicates the commonly used investment casting instrument with its corresponding wax component shells, shown in **Fig. 3.3**. In addition, some finished products are shown in **Fig. 3.4**. **Table 3.1** lists some benefits and shortcomings of IC.

**FIG. 3.2** Investment casting equipment

**FIG. 3.3** Some wax components of investment casting

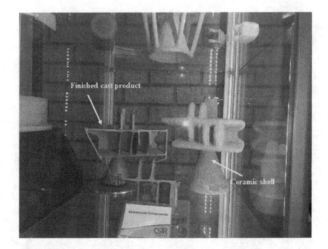

**FIG. 3.4**   Some commercial components

**TABLE 3.1**
**Advantages and disadvantages of the IC process**

| Advantages | Disadvantages |
| --- | --- |
| • Investment casting is ideal for complex, intricate-shaped parts, hence, it has the freedom of design. All materials can be produced. It is mostly used for All alloy, cast iron, and non-ferrous alloys, particularly attractive for high-temperature alloys.<br>• The lost wax provides superior surface in finishing and outperforms other casting processes with precise tolerances better than sand casting, forging, or fabrications.<br>• Good tolerance level with little machining required.<br>• It produces castings with good quality and fewer casting defects.<br>• Due to the near-net-shaped products little machining minimizes material waste and requires less expensive equipment, and it is a safe process.<br>• It accommodates production of large and small casts.<br>• Variety of materials can be used for investment casting such as carbon steel, alloy steel, heat resistant alloy | • It has a size limitation.<br>• This process is best for casting small intricate components. Excessively large castings are not suitable.<br>• Investment casting shells have limitations on size and depth. The size cannot be smaller than 1.6 mm or deeper than 1.5 times the diameter.<br>• The process requires a substantial amount of preparation and specialized equipment. It can be more expensive than sand casting or die casting, but high volumes mitigate the high cost. |

## 3.3 VACUUM ARC MELTING

Vacuum arc remelting (VAR) is a secondary metallurgical process used to manufacture high-purity steel, and nickel-based, and titanium-based alloys. The process involves heat supplied to a remelting electrode through a direct current (DC) arc. On the other hand, the electrode surface exhibits cathode spot development and mobility, a vacuum arc, and direct current transmission between the electrode and mould [1].

The vacuum arc melting (VAM) equipment is mainly used for melting of metals by electric arc placed between the crucible containing metal and a tungsten electrode in a copper hearth as shown in **Fig. 3.5** below, which is also an illustration of combining both powder and ingot metallurgy routes to produce buttons. The chamber is filled with argon (Ar) and melting is conducted in Ar. **Fig. 3.6** is vacuum equipment used for the production of bulk components.

A Tungsten Inert Gas (TIG) welding unit is the power source and the heat generated by the electric arc melts the metals placed in the crucible to form a mixture of an alloy. In most instances, the melting is repeated to produce a homogeneous alloy. Oxidation is avoided by evacuating the chamber while the metal cannot react with the inert Ar.

The metals can be heated to a temperature more than 2000 °C. A batch of five alloys can be made in a single evacuation, as there are five crucibles in the hearth (four small and one large). About 15g of metals can be melted in the small crucibles and about 80g in the larger crucible. There are three main parts to the system: power source (TIG– 600Amp), chiller, and vacuum unit. The vacuum unit with rotary and diffusion pumps can attain a vacuum of $10^{-6}$ mbar. The cold circulation water from the chiller cools both the copper hearth and the electrodes. After the elemental metals (or master alloy) are melted and solidified, they can be 'turned over' by a 'tweezer mechanism' without breaking the vacuum (and then re-melted). The melting →

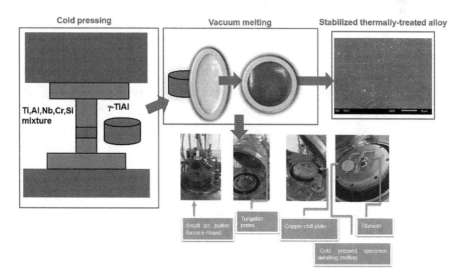

**FIG. 3.5** A schematic illustration of the small vacuum button arc melting, of γ-TiAl alloy doped with Cr, Nb, and Si elements

**FIG. 3.6**   Vacuum equipment for bulk processing

solidification → 'turn over' of sample → remelting process is typically repeated three times to attain a better compositional homogeneity. Apart from the abovementioned hearth with five crucibles, an additional hearth has been provided with one crucible, which can suction cast the molten alloy, in the form of thin cylinders (typically 3mm diameter).

Some studies showed that the VAM processed alloys have higher wear resistance compared to their cast counterparts owing to the smaller surface area [2]. Moreover, the VAM has led to superior corrosion performance as compared to stir casting and sintering routes [2, 3].

## 3.4   CASTING PROCESS IMPROVEMENT EMPLOYING NUMERICAL TOOLS AND EXPERIMENTAL TEST

A population of defects in components are due to inaccuracies in casting design. In the old foundries or metal casting organizations, the changes to design were done by using a trial-and-error method which increases costs in every step of the design and manufacturing process. Foundries relying on traditional trial-and-error approach are finding increasingly difficult to compete in the production of high-quality cast products and efficiency in productivity. Casting modelling and simulation has become an essential tool in optimization of casting parameters, design for manufacture, and tool and pattern design. In nature, casting simulation is a complex multiphysics simulation, which combines computational fluid dynamics (CFD), heat and mass transfer, solidification, and metallurgy to predict the quality of the component produced for a particular casting process. It is used to simulate and predict the quality of the casting after solidification. The simulation software can be used to optimize:

- casting design
- tool and mould design
- processing parameters

Major benefits of using casting simulation are:

- improved product quality
- design for manufacture – reduce lead time for product development
- reduction in lead time and costs for tool and mould designs
- reduction in rework of tools which can add costs and extend lead times, because permanent mould casting tools are very expensive
- mould design for sand and investment casting
- fewer process changes, better success rate and reduction of scrap
- can be used for predicting real costs to make a casting

### 3.4.1 CASTING SIMULATION SOFTWARE

There are a number of casting simulation software packages available commercially. The two most widely used packages are ProCAST, supplied by ESI Group as a module in their multiphysics suite of software, and MAGMA, supplied by Magmasoft.

The casting simulation capability serves multiple casting platforms such as high-pressure die casting, gravity die casting, sand casting, and investment casting.

### 3.4.2 DEVELOPMENT

Casting simulations can improve the gating system and optimize the design positively. Zhizhong [4] optimized the gating system parameters using the Taguchi method in combination with advanced numerical simulation. The Taguchi method obtains a set of optimal gating system parameters based on the defined objectives. As a result, Taguchi's orthogonal array provides a suitable and efficient methodology for the optimization of gating system parameters. The product yield, shrinkage porosity, and filling velocity can be simultaneously considered and improved through this optimization technique.

Moreover, SOLIDWORKS and ProCAST software as a modelling tool and a casting numerical simulation tool was used based on structural analysis of a water-meter shell, a three-dimensional model, and a finite element model. Three processes accomplished by the bottom gating system without a riser, a step gating system with a preliminary riser, and a step gating system with an optimum riser were numerically simulated. The optimization of the casting process and the rational assembling of the riser led to the shrinkage volumes at the inlet position, regulating sleeve, and sealing ring of the water-meter shell decreasing from 0.68 to 0 cm$^3$, 1.39 to 0.22 cm$^3$, and 1.32 to 0.23 cm$^3$ [5]. The model predictions and experimental measurements showed that castings produced by the optimized process had good surface quality and appearance, and were defect-free, validating the effectiveness of numerical simulation for improving casting quality.

## 3.5    DESIGN AND MANUFACTURE OF INTRICATE PROSTHESES

A **prosthetic implant** is an artificial device made from any biofriendly material with good properties that replaces a body part that has been removed due to failure or deteriorating conditions because of trauma, disease, or a condition present at birth (congenital disorder). They restore the normal functions of the missing body part, normally coordinated by a medical team with individual experts on different health specialisms. Prostheses can be manufactured by hand or with computer-aided design (CAD), a software interface that helps creators design and analyze the creation with computer-generated 2-D and 3-D graphics as well as analysis and optimization tools. For example, prostheses used for replacement of the hand can be divided into active and passive prostheses. The force to control the grasping mechanism of active prostheses is internal by an electric actuator or a body-powered cable. However, in passive prostheses, the force to adjust the grasping mechanism is external by the sound hand [6].

A prosthesis should be designed and assembled according to the person's appearance and functional needs. Craniofacial prostheses include intra-oral and extra-oral prostheses, while the extra-oral prostheses are divided into hemifacial, auricular (ear), orbital, nasal, and ocular. Intra-oral prostheses include dental prostheses, such as obturators, dentures, and dental implants. Larynx substitutes and trachea and upper esophageal are prostheses of the neck.

Somato prostheses of the torso include breast prostheses which may be either single or bilateral, full breast devices, or nipple prostheses. Penile prostheses are used to treat erectile dysfunction, correct penile deformity, perform phalloplasty and metoidioplasty procedures in biological men, and to build a new penis in female-to-male gender reassignment surgeries.

Limb prostheses are upper- and lower-extremity prostheses. Upper-extremity prostheses are used at varying levels of amputation: forequarter, shoulder disarticulation, transhumeral prosthesis, elbow disarticulation, transradial prosthesis, wrist disarticulation, full hand, partial hand, finger, and partial finger. An artificial limb that replaces an arm missing below the elbow is called a transradial prosthesis.

### 3.5.1    UPPER LIMB PROSTHESES

There are three main categories:

(a) Passive devices. Passive devices can either be passive hands, mainly used for cosmetic purposes, or passive tools, mainly used for specific activities (e.g. leisure or vocational). An extensive overview and classification of passive devices can be found in a literature review by **Maat et.al.** [6]. A passive device can be static, meaning the device has no movable parts, or it can be adjustable, meaning its configuration can be adjusted (e.g. adjustable hand opening). Despite the absence of active grasping, passive devices are very useful in bimanual tasks that require fixation or support of an object, or for gesticulation in social interaction. According to scientific data a third of the upper limb amputees worldwide use a passive prosthetic hand [6].

(b) Body-powered devices. Body-powered or cable-operated limbs work by attaching a harness and cable around the opposite shoulder to the damaged arm.

(c) Externally Powered (myoelectric) devices. Myoelectric arms work by sensing, via electrodes, when the muscles in the upper arm move, causing an artificial hand to open or close. In the prosthetics industry, a transradial prosthetic arm is often referred to as a 'BE' or below elbow prosthesis.

### 3.5.2 LOWER LIMB PROSTHESES

Lower-extremity prosthetics describe artificially replaced limbs located at the hip level or lower and provide replacements at varying levels of amputation. These include hip disarticulation, transfemoral prosthesis, knee disarticulation, transtibial prosthesis, Syme's amputation, foot, partial foot, and toe. The two main subcategories of lower-extremity prosthetic devices are transtibial (any amputation transecting the tibia bone or a congenital anomaly resulting in a tibial deficiency) and transfemoral (any amputation transecting the femur bone or a congenital anomaly resulting in a femoral deficiency) [6].

A transfemoral prosthesis is an artificial limb that replaces a leg missing above the knee. Transfemoral amputees can have a very difficult time regaining normal movement. In general, a transfemoral amputee must use approximately 80% more energy to walk than a person with two whole legs [6]. This is due to the complexities in movement associated with the knee. In newer and more improved designs, hydraulics, carbon fiber, mechanical linkages, motors, computer microprocessors, and innovative combinations of these technologies are employed to give more control to the user. In the prosthetics industry, a transfemoral prosthetic leg is often referred to as an 'AK' or above the knee prosthesis.

A transtibial prosthesis is an artificial limb that replaces a leg missing below the knee. A transtibial amputee is usually able to regain normal movement more readily than someone with a transfemoral amputation, due in large part to retaining the knee, which allows for easier movement. In the prosthetics industry, a transtibial prosthetic leg is often referred to as a 'BK' or below the knee prosthesis.

Physical therapists are trained to teach a person to walk with a leg prosthesis. To do so, the physical therapist may provide verbal instructions and may also help guide the person using touch or tactile cues. This may be done in a clinic or home. There is some research suggesting that such training in the home may be more successful if the treatment includes the use of a treadmill [6]. Using a treadmill, along with the physical therapy treatment, helps the person to experience many of the challenges of walking with a prosthesis.

In the United Kingdom, 75% of lower limb amputations are performed due to inadequate circulation (dysvascularity) [6]. This condition is often associated with many other medical conditions (co-morbidities) including diabetes and heart disease that may make it a challenge to recover and use a prosthetic limb to regain mobility and independence [6]. For people who have inadequate circulation and have lost a lower limb, there is insufficient evidence, due to a lack of research, to inform them regarding their choice of prosthetic rehabilitation approaches [6]

A typical above-knee prosthesis has a foam layer which is generally covered with artificial skin that is painted to match the patient's natural skin colour.[7].

Lower-extremity prostheses are often categorized by the level of amputation or after the name of a surgeon:[8]

- Transfemoral (above-knee)
- Transtibial (below-knee)
- Ankle disarticulation (e.g.: Syme amputation)
- Knee disarticulation
- Hemi-pelvectomy (hip disarticulation)
- Partial foot amputations (Pirogoff, talo-Navicular and calcaneo-cuboid (Chopart), tarso-metatarsal (Lisfranc), trans-metatarsal, metatarsal-phalangeal, Ray amputations, toe amputations) [8]
- Van Nes rotationplasty

### 3.5.3 PROSTHETIC RAW MATERIALS

Prosthetic are made lightweight for better convenience for the amputee. Some of these materials include:

- Plastics:
  - polyethylene
  - polypropylene
  - acrylics
  - polyurethane
- Wood (early prosthetics)
- Rubber (early prosthetics)
- Lightweight metals:
  - titanium
  - aluminium
- Composites:
  - carbon fiber reinforced polymers[4]

Wheeled prostheses have also been used extensively in the rehabilitation of injured domestic animals, including dogs, cats, pigs, rabbits, and turtles.

## 3.6 CONCLUSIONS

This chapter gives an overview of the ingot metallurgy routes focusing on metal casting but taking an example of investment casting and vacuum arc melting. It shows that with the latest technological developments, modelling and simulation is crucial for product development. As a result, very difficult castings can be made with fewer defects and at a minimum cost.

# REFERENCES

[1] E. Karimi-Sibaki, A. Kharicha, M. Wu, A. Ludwig, & J. Bohacek, A Parametric study of the vacuum arc remelting (VAR) process: Effects of arc radius, side-arcing, and gas cooling, *Metallurgical and Materials Transactions B*, 51B, 222, 2020.

[2] A. G. Lekatou, A. K. Sfikas, & A. E. Karantzalis, The influence of the fabrication route on the microstructure and surface degradation properties of Al reinforced by Al9Co2, *Mater. Chem. Phys.* 200, 33–49, 2017.

[3] A. Lekatou, A. K. Sfikas, A. E. Karantzalis, & D. Sioulas, Microstructure and corrosion performance of Al-32%Co alloys, *Corrosion Sci.*, 63, 193–209, 2012.

[4] S, Zhizhong, Numerical optimization of gating systems for light metals sand castings, *Electronic Theses and Dissertations*, 2886, 2008. https://scholar.uwindsor.ca/etd/2886

[5] K. Zheng, Y. Lin, W. Chen, & L. Liu, Numerical simulation, and optimization of casting process of copper alloy water-meter shell, *Advances in Mechanical Engineering*, 12, (5), 1–12, 2020, DOI:10.1177/1687814020923450, journals.sagepub.com/home/ade.

[6] B. Maat, G. Smit, D. Plettenburg, & P. Breedveld, Passive prosthetic hands and tools: A literature review, Prosthetics *and Orthotics International*, 42 (1), 66–74. 2018.

[7] A. M. Turing & J. von Fraunhofer, How artificial limb is made – material, manufacture, making, used, parts, components, structure, procedure. www.madehow.com. Retrieved 2022-09-05.

[8] S. Bengt, *Partial foot amputations* (2nd ed.). Sweden: Centre for Partial Foot Amputees. 2001. ISBN 978-9163107566. OCLC 152577368.

# 4 Powder-Based Manufacturing Techniques
## Powder Metallurgy Routes

*A. S. Bolokang and M. N Mathabathe*

## 4.1 INTRODUCTION

Powder metallurgy (PM) is a manufacturing process used for producing cost-effective materials components for a range of engineering applications. It is a method capable of producing products which are complicated to produce by wrought processing techniques. The PM advantage over other techniques is to produce near-net-shape and homogeneous parts with reduced machining and scrap rate. It is a solid state which can be used in batch production systems. Applications for PM include automotive, biomedical devices and mining. The PM process involves:

(1) mixing metal powders including reinforcement powders where composite and metal-matrix composites are made
(2) compaction of the prepared composition into metal dies to produce green materials
(3) Annealing or sintering of the compacts and further processing, i.e., forging, extrusion etc.

Alternatively, metal injection moulding (MIM) and additive manufacturing (AM) are well-established PM processes. These techniques have gained a considerable market share in the medical devices market. It is critical to tailor the mechanical properties in PM manufacturing to gain advantage over conventional manufacturing relevant to specific patients. For example, porous biomedical products can be tailored to address particular medical needs of patients. Additionally, powder rolling (PR) enables the manufacturing of thin-walled sheets/plates.

Moreover, $Ni_{62.5}Al_{37.5}TiC_{1.28}$ composite was developed by powder mixing, compaction, and sintering at 650 °C. The chemical reaction during sintering showed that thermal explosion (TE) occurred when a small amount of nanosized TiC powder was added. The effect of cold pressing (CP) on low-temperature TE is proposed. Martensite NiAl laths, $Ni_3Al$ and TiC phases were formed.

DOI: 10.1201/9781003356714-4

### 4.1.1  RAW MATERIALS: ELEMENTAL METALLIC POWDERS

The critical component of PM is the raw material. Powders are manufactured through different processes, and they can be elemental powders or alloy powders.

Elemental powders serve as raw materials to alloy powder mixtures. Characterization and quality control of powder is crucial for alloy development. This includes the purity levels, morphology, and thermal properties of elemental powders. The different methods to produce powders are:

(1) Atomization: this process produces both ferrous and non-ferrous metallic powder alloys such as shape memory alloys, superalloys, and stainless steel, while elemental include iron (Fe), copper (Cu), molybdenum (Mo), titanium (Ti) etc.

(2) Electrolysis: this process is used to make Fe, Cu, cobalt (Co), silver (Ag), nickel (Ni) etc.

Typical morphologies of powders are determined by the process used to manufacture different metal elemental powders. **Fig. 4.1** shows Ti powders displaying both the spherical and irregular particle morphology. Elemental powders reveal different properties, for example, the Ni, Fe, and Co are ferromagnetic while metals such as Ti, W, Cu etc. are non-magnetic in nature.

**Fig. 4.1–4.2** show the Ti powder resembling particle shapes produced from the atomization and chemical processes, respectively. Gas atomization process produced the spherical Ti particles while the chemical process yielded angular-shaped powder particles.

Both the Cu and Ni powders also reveal different morphologies (**Figs. 4.3–4.4**). The ultrafine spherical Ni powder formed chains of small particles. The physical properties, particle sizes, and shapes of elemental powders contribute to the formation during compaction of either unmixed or mixed powders. The elemental constants and properties are presented in [1].

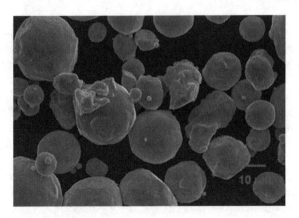

**FIG. 4.1**  SEM image of the atomized Ti powder

**FIG. 4.2** SEM image of the Ti powder produced by chemical processes

**FIG. 4.3** SEM image of the Cu powder

Both Ni and Co are ferromagnetic materials and produce a curie temperature, the second-order phase transitions. The phase transitions indicate the magnetic transition from ferro-to-paramagnetic state. With Co being HCP, it also transforms from HCP to FCC at high temperature like Fe. Cu is an FCC metal which doesn't show any phase transition. On the other hand, Ti shows an allotropic phase transformation from HCP to BCC at high temperature. This phase transformation occurring in Ti, Co, and Ni are reversible at low temperature and affected by alloying, mechanical alloying, and forming processes.

**FIG. 4.4**   SEM image of the Ni powder

## 4.2   ADDITIVE MANUFACTURING

AM as a process has been around since the late 1980s. This process uses computer-aided design (CAD) and 3D printing software to guide digital hardware to accurately produce detailed geometric shapes, layer-by-layer deposition. By utilizing CAD or a 3D scanning machine, the creation of products with intricate and accurate geometric shapes is accomplished. The difference between 3D printing and other processes is that no machining or fettling of the component is required to meet dimensional tolerances.

### 4.2.1   BENEFITS OF ADDITIVE MANUFACTURING

There are different types of AM processes (**Table 4.1**), and their benefits include a certain degree of waste reduction and/or energy savings. **Table 4.2** outlines the advantages and disadvantages of AM.

### 4.2.2   APPLICATIONS OF ADDITIVE MANUFACTURING

#### Consumer products

AM is beneficial to the product development of many consumer goods such as sporting goods and consumer electronics. The detailed iterations can be delivered

**TABLE 4.1**
**Types of manufacturing processes**

| Types of AM processes | Description |
| --- | --- |
| Binder jetting | 3D printing or powder bed and inkjet. It uses liquid materials printed onto thin layers of powder. It builds layer by layer and binds the particles together. Typical examples are foundry sand, metals, and ceramics in construction of low-cost 3D printed metal products and large sand-casting moulds. |
| Power bed fusion | This process includes the various printing techniques of selective laser sintering and melting, direct metal laser sintering, electron beam melting, and selective heat sintering. The high-power thermal energy through laser or electron beam, melts and bonds material powder together. |
| Material jetting | The deposition of material's droplets is selectively but continuously deposited onto the build platform. Curing occurs by a heat source or ultraviolet light to form a 3D object. |
| Material extrusion | This is a fused layer modelling or fused filament fabrication; material extrusion uses continuous filament to construct 3D parts. The material is extruded through a nozzle, heated, and deposited, layer by layer, onto the build platform. |
| Directed energy deposition | It is a process using thermal energy to fuse metal and metal-based materials by melting the material as it is deposited. It is laser metal deposition, plasma arc melting, and direct metal deposition used for low volume part production, rapid prototyping, and repair. |
| Vat photopolymerization | This process is a type of additive manufacturing technology that produces 3D products by selectively curing photopolymer liquid resin using light-activated polymerization. A photopolymer is a polymer that changes its properties when exposed to light, often in the ultraviolet or visible region of the electromagnetic spectrum, causing its molecules chain to link. |
| Sheet lamination | Sheet lamination bonds together thin sheets of material. The sheets are usually fed via a system of rollers and bonded together, layer by layer, to form a single sheet. |

quickly in the early product development life cycle, with fine details, functionality, and realistic aesthetics.

## Energy

AM applications in the gas, oil, and energy industries include various control-valve components, turbine nozzles, rotors, flow meter parts, pressure gauge pieces, and pump manifolds.

**TABLE 4.2**
**Advantages and disadvantages of additive manufacturing**

| Advantages | Disadvantages |
| --- | --- |
| • Accelerated prototyping, faster and cheaper in comparison to lengthy traditional methods. Prototypes can be manufactured prior to real products. | • The capital costs to buy equipment to replace investment made on the plant and equipment for their traditional operations is still substantial. |
| • The process offers design innovation and creative freedom without the cost and time constraints of traditional manufacturing | • It is a slow process and still a niche process. The build-up rates are slow, therefore, high volume production is a serious limitation. |
| • AM uses fewer resources, less need for ancillary equipment, and reduces waste material. | • Production costs are high because specific materials for AM are frequently required in fine or small particles that can considerably increase the raw material cost of a project. The surface finish is not always good. |
| • With AM, products are printed on demand, there is no over-production. Production is made for sale. | |
| • The legacy parts difficult and expensive to produce can now be produced through the scanning and X-ray analysis of original material and parts. | • A post-processing is required in AM because surface finishes and dimensional accuracy can be of a lower quality compared to other manufacturing methods. |
| • Manufacturing can be combined and assembled into a single process to consolidate manufacture and assembly. | |
| • AM enables the design and creation of many geometric forms that reduce the weight of an object while still maintaining stability. | • Limitations on the types of materials, typically pre-alloy materials in a base powder. |
| • AM reduces the number of component defects on intricate parts | • The mechanical property of a finished product depends on the quality of the powder used in the process. |
| • Flexible production rates. A key benefit of AM is the ability of producers to quickly switch between different products | • Small moving pieces typically require strict manufacturing tolerances and very controlled assembly process. |
| • Improves supply chain. AM improves process flexibility that enables supply chains to quickly react to demand; lowers supply chain risk by providing a contingency plan and helps to reduce supply-related costs. | |

## Transportation

Parts that withstand extreme speeds and heat, but are lightweight enough to avoid preventable drag, are achieved. The benefit of AM's ability to develop lightweight components has led to more efficient vehicles.

## Aerospace

The aerospace industry has adopted AM. This includes environmental control system ducting, custom cosmetic aircraft interior components, rocket engine components, and combustor liners. AM is useful in complex, consolidated parts with the enhanced strength that is a requisite in the industry.

## Medical

Material development in the medical industry is biocompatible body implants, and the life-saving devices and pre-surgical tools to improve patient outcomes. Orthopaedic implants and dental devices as well as tools and instrumentation such as seamless medical carts, anatomical models, custom saw and drill guides, and custom surgical tools are realized.

## 4.3   UNIAXIAL COLD PRESSING AND SINTERING

The compacts for determining green density and strength are obtained using a high-speed tool steel cylindrical die (**Fig. 4.5**) of varying diameters. For optimum green strength, the masses of powders are chosen to ensure that the aspect ratio of thickness, $t$, to diameter, $D$, is less than 0.25, as stipulated in ASTM D3967. De la Torre et. al., [2] repeated the cold pressing on two pieces of metal (Sn and Te) and induced alloying on metals. By using XRD and DSC methods, it was established that the alloying may be on surface particles. This implies that the crystal orientation has changed. Apart from the alloying of two or more metals, phase and structural transformation has

**FIG. 4.5**   Enerpac 100T uniaxial compaction machine

been confirmed in pure metals like Co upon cold pressing, also shown in Karin et al. [3]. However, some metals are easier to deform than others. It is logical that uniaxial pressing stresses occur during the sintering process of compacted powders. **Fig. 4.6** shows the uniaxial cold-pressing mechanism of a binary mixture of titanium and nickel powder (Ti 50 at.%Ni). The powder mixture is composed of different particle sizes and shapes. The Ti particles are mostly smooth spherical gas atomized produced and larger while the fine Ni particles have a chain structure produced by chemical process (**Figs. 4.1** and **4.4**) respectively.

In the uniaxial cold-pressing process, the powder is exposed to external applied pressure as illustrated in **Fig. 4.6**. A green compact with porosity is produced. In the case of elemental powder, the property of the powder determines if a lubricant such as stearic acid is required to ease the compatibility of the powder. Some powders require extensive pressure to yield a compact while others require a small amount of pressure to bond particles together.

For alloys, powders are mixed with additives or lubricants such as stearic acid, also using mixers for easy removal of the compact and to avoid wear of tools. The loose powder particles are formed into the appropriate shapes, yet with enough strength to withstand the sintering temperatures that will result in full bonding and formation, without the use of heat. The hot compaction can be implemented on cemented carbides prior to the final sintering process. The loose powder compaction can be accomplished without pressure application by sintering loose powder in a mould, vibratory compaction, injection moulding, slip casting etc. When pressure is applied, cold die compaction by single or double pressing, isostatic pressing, explosive compacting, and powder rolling are applicable. After cold pressing, high density near-net shape green parts are produced. **Fig. 4.6** shows a schematic illustration of the cold-pressing process. The pressure is applied simultaneously and equally in all

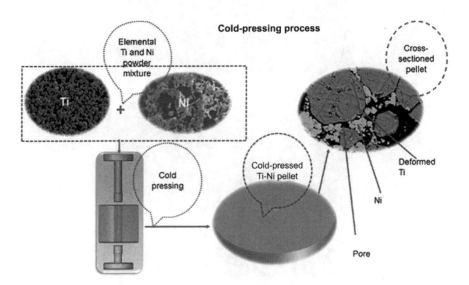

**FIG. 4.6**   Uniaxial cold-pressing process to produce a green powder compact

**FIG. 4.7**   Gas Pycnometer (AccuPyc II 1340)

directions on the coarse Ti and ultrafine Ni powder mixture. The ultrafine Ni powder is distributed around the Ti spherical powder but spread towards the die surfaces. The compacted green parts can be compacted up to 70–90% of their theoretical densities depending on the applied pressure [3]. Density was measured using Pycnometer **Fig. 4.7**. The porosity is eminent after compaction to be minimized after sintering during chemical reaction. Metals have different physical properties as illustrated in **Table 4.1**. For example, Ti has a yield strength of 200 MPa while Ni has 130 MPa. The Ni metallic powder is compacted better compared to Ti. In the TiNi alloy mixture, the Ni powder acts as a binder to Ti particles because it is easily deformable. The sintering process occurs by diffusion between Ni and neighbouring Ti particles. The resultant TiNi alloy is a specific chemical composition that adopts new mechanical, chemical and physical constants. Both parent and daughter metal contribute to the final density, modulus of elasticity (MOE), yield strength, thermal expansion, melting temperature, and crystal structure. HCP Ti and FCC Ni may yield a monoclinic/body-centred cubic TiNi alloy subject to the chemical composition.

Thermal behaviour of the compacted Ti-4.5Ni binary alloy is different from the thermal behaviour of pure Ti. An endothermic $\alpha \rightarrow \beta$ Ti phase transformation peak occurs at 773 °C compared to 882 °C in pure Ti. The 4.5 wt.%Ni content affects the $\alpha \rightarrow \beta$ phase transformation temperature of Ti [4].

### 4.3.1   SURFACE PROPERTIES AND COMPACTION

The spherical Ti powder builds an oxynitride layer due to annealing the powder in air forming $TiO_{2-x}$-$N_x$ particles. Annealing at 500, 600, 700 and 800 °C confirms the particle surface oxidation as a function of temperature by different colours. The $TiO_{2-x}$-$N_x$ powder were mixed with Al, Sn, and Ni and compacted at 200 MPa. The $TiO_{2-x}$-$N_x$ powder is difficult to compact but in a mixture with low yield strength Al, Sn, and Ni, compaction is achieved. A high-resolution scanning electron microscope

(HR-SEM, Auriga ZEISS) coupled with a Robinson Backscatter Electron Detector and an Oxford Link Pentafet energy dispersive x-ray spectroscopy (EDS) detector reveals the morphology of the polished composite in **Fig. 4.8**. The compacts were sintered at 800 °C. At 800 °C, $TiO_{2-x}N_x$ layer formed on the surface of the spherical Ti particles [5]. They are potential additives for metal-matrix composites (MMCs) parts.

The Ti particle has higher affinity to O than N forming $TiO_2$ oxide which dissolved the N by $2N^{-3}$ occupying a $2O^{-2}$ position, resulting in a $TiO_{2-x}N_x$ oxynitride layer [5]. The diffusion coefficient (*D*) and temperature (*T*) is correlated using the Arrhenius **Eq. 4.1**:

$$D = D_0 \exp\left(\frac{-Q}{RT}\right)\pi r^2 \qquad (4.1)$$

where $D_0$ is the exponential factor, *R* is the gas constant (8.31 J mol−1 K), *Q* is the activation energy and *T* is the absolute temperature.

**Fig. 4.8** shows the sintered MMC revealing the hard spherical $TiO_{2-x}N_x$ particles with dark surface coating due to O and N dissolution. Some particles were subjected to severe oxidation and collapsed as a result of different particle sizes, which started the porosity. An Ni, Al, Sn matrix wets the ceramic particles, see **Fig. 4.9**. The chemical analysis of the sintered cold-pressed sample is shown in **Fig. 4.8**.

**FIG. 4.8**    Optical and SEM images of the 800 °C sintered Al-1Sn-5Ni-(TiO2-xNx) composites

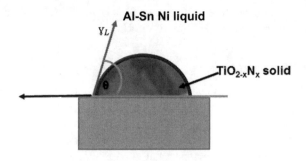

**FIG. 4.9**    Illustration of the solid $TiO_2$-xNx particle wetting by Al-1Sn-5Ni liquid metal

At a sintering temperature of 800 °C, Al-1Sn-5Ni acts as a binder to provide liquid transport for wetting of the solid $TiO_{2-x}N_x$ spherical particles. It also provides a surface tension force assisting in densification of the material. A wetting liquid A356-1Sn-5Ni metal makes a small contact angle with solid $TiO_{2-x}N_x$ particles defined by **Eq. 4.2** and illustrated in **Fig. 4.6**:

$$\gamma_{SV} = \gamma_{SL} + \gamma_{LV} Cos\,\theta \qquad (4.2)$$

where $\gamma_{sv}$ represents solid-vapour surface energy, $\gamma_{SL}$ is solid-liquid surface energy and $\theta$ indicates that liquid will cover the solid particle [5].

**Fig. 4.10** represents the EDS elemental mapping confirming the presence of Al, Sn, Ti, and Ni. The spherical solid particles are classified as Ti. The analysis of O and

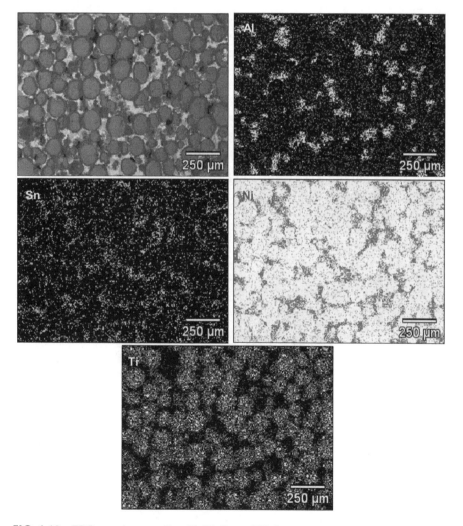

**FIG. 4.10**   EDS mapping revealing Al, Ni, Sn, and Ti elements

N was reported in the previous work to produce the $TiO_{2-x}N_x$ spherical particles [5] and can form by simultaneous O and N diffusion during annealing in air [6].

## Case study 1: Cold-pressed and sintered $Ni_{62.5}Al_{37.5}TiC_{1.28}$ composite

### Nickel aluminides

The nickel aluminide (NiAl) alloys are interesting systems with a melting temperature of 1640 °C and contain the ordered B2 intermetallic NiAl with unique physical and mechanical properties for high-temperature applications [7]. However, there is a need to improve its low-temperature strength which limits the alloy application. The NiAl system reveals a range of phases depending on the starting chemical composition. These phases include the high-temperature $Al_3Ni$ eutectic at elevated temperatures used in production of aerospace gas turbine engine blades, hence it is considered as an attractive structural material for high-temperature applications [8]. In particular, the Al45 at.%Ni formed via explosion of electrically heated twisted pure Al and Ni wires in an argon atmosphere [9]. Operating temperatures as low as 575 °C–750 °C are effective to synthesize a porous NiAl intermetallic alloy due to exothermic reactions [10]. TE and ignition of various NiAl powder blends, rolled and magnetron sputtered films, and compacts of mechanically activated and non-activated mixtures were explored under mechanical activation. A single cubic B2–NiAl phase was synthesized through mixing of elemental Ni and Al powders followed by cold compacting and sintering [11]. Cubic B2–NiAl alloy was synthesized after sintering of the cold-pressed compacts at 750 °C and 1300 °C, respectively. The compacts sintered at 1300 °C were brittle while those sintered at 750 °C did not show the brittle behaviour. The TE process of a stoichiometric NiAl system can be investigated by time-resolved X-ray diffraction techniques to trace the dynamics of phase transformation during rapid, gasless heterogeneous reactions [12]. The TE of mechanically milled and unmilled NiAl is compared, with the former occurring owing to melting and the latter due to solid-state reaction. This behaviour implies that deformations create fresh surfaces for a solid-state reaction to occur between Ni and Al particles during mechanical milling. The mechanical activation leads to a decrease in the activation barrier while ignition becomes possible at low temperatures below the melting point of Al [13].

### The $Ni_{62.5}Al_{37.5}TiC_{1.28}$ composite

Nanosized TiC powder was added to an Ni and Al composition. Nickel becomes a strategic metal for both ferromagnetic and non-magnetic shape memory alloys (SMAs) [14–16]. Despite limited investigations on this metal's properties linked to SMAs, it has been shown that it exhibits the second-order phase transition (ferromagnetic to paramagnetic) when heated to high temperatures [17, 18]. This phase transition is affected by mechanical deformations and impurities [19, 20]. The NiAl alloy forms martensite and the nanosized TiC induces high strength. The cold pressing influences the alloy during sintering. Elemental Ni (99.8%), Al (99.9%), and nanoparticles of TiC powders are blended according to an $Ni_{62.5}Al_{37.5}TiC_{1.28}$ composition. The powder mixture undergoes cold pressing to form compacts of either products or laboratory-scale samples. Compactibility and strength of the green samples depends on the pressure. This composite was at 650 °C in a Carbolite tube furnace flowing with argon (Ar) for 4 hours (h). The high-resolution scanning electron microscope (HR-SEM, Auriga ZEISS) coupled with a Robinson Backscatter Electron Detector and an Oxford Link Pentafet (EDS) detector revealed the composite microstructure and powder morphology. MicroVickers hardness tests were performed applying a load of 300 Kgf and dwelling time of 10 seconds (s) were employed for the microhardness measurements. Hardness profiles throughout the specimens were measured at 0.2 mm intervals and average of 32 measurements.

## Results and discussions

The SEM images of the pure Ni powder reveals chains of ultrafine, spherical particles (~2 μm) revealing the agglomeration as shown in **Fig. 4.11**. The ultrafine particles are mainly attached to each other.

A rough stonelike Al powder morphology is shown in **Fig. 4.12** reveals an agglomeration of particles that formed large particles ~40 μm in size. Due to the low yield strength of the Al when compared with both Ni and TiC particles, its particles respond quickly to the pressure applied during cold pressing. Al then lubricates Ni and TiC particles. The TiC powder also appears fine and agglomerated as shown in **Fig. 4.13**. Cold pressing influences particle interaction between Ni and Al particles since the inter-particle contact makes the thermal reaction and ignition of the aggressive exothermic reaction. The thermodynamic reactions 1 and 2 occur below the melting temperature of pure Al (660 °C) since the sintering process was performed at 650 °C. However, although the melting temperature of Ni is 1453 °C, Ni undergoes the second-order phase transformation at ~356 °C (Curie point) becoming paramagnetic [**17, 18**].

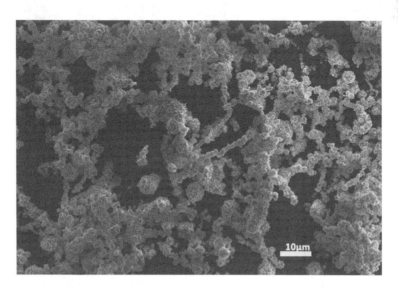

**FIG. 4.11**   SEM images of the Ni, Al, and TiC powders

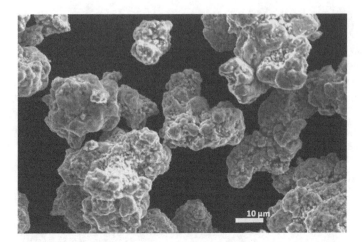

**FIG. 4.12**   SEM image of the Al powder

**FIG. 4.13**   SEM image of the TiC powder

This transition temperature can be affected by deformation of the powder particles. The deformation induced on the particle surfaces after cold pressing contributes to the ignition temperature occurring below the melting temperature of Al [**10, 19**], also leading to a decrease in the activation barrier [**21**]. **Fig. 4.14b, c** reveals the thermal analysis of the cold-pressed and free-flowing Al powders. It is evident that the cold-pressed sample melted ~10 °C lower than the free-flowing powder. This behaviour agrees with our proposed mechanism on the effect of cold pressing on the ignition

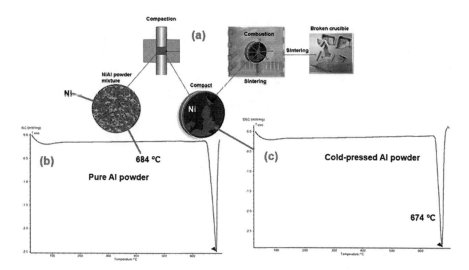

**FIG. 4.14**   Schematic representation of the synthesis and thermal explosion of $Ni_{62.5}Al_{37.5}$ $TiC_{1.28}$ composition during sintering

temperature of NiAl compacts during sintering. As a result, the occurrence of TE is justified by the broken crucible during sintering. The samples were all found attached to the bottom of one piece of the broken crucible, indicative of the expansion of the compacts during sintering.

$$Ni + 3Al = NiAl_3 \tag{4.3}$$

$$1/3Ni_2Al_3 + 1/3Ni = NiAl \tag{4.4}$$

**Fig. 4.15a–d** shows the SEM images of the $Ni_{62.5}Al_{37.5}TiC_{1.28}$ composite sintered at 650 °C. The exothermic reaction accompanied by the TE during sintering is attributed to the alloy formation. **Fig. 4.15a** shows the resulted microstructure revealing all phases present after sintering. Of interest in this study, the fine martensite NiAl laths formed are shown in **Fig. 4.15c**. The martensite plates are a ductile structure surrounded by the eutectic-type $Ni_3Al$ structure and hard TiC phase. The $Ni_3Al$ phase shows the martensite plates when rapidly cooled bcc B2 β-phase to f.c.t. $L_{10}$ [22]. These samples were furnace cooled. The TiC grains (**Fig. 4.15b, d**) reveal the sharp-edged and faceted structures. This $Ni_{62.5}Al_{37.5}TiC_{1.28}$ composite resists penetration by the diamond-cutting tool which is due to the ultrahard TiC presence. TiC particles embedded inside the $Ni_3Al$ martensite plates inhibit deformation. Therefore, reinforcement of the $Ni_{62.5}Al_{37.5}$ intermetallic alloy by nanosized TiC particles of 1.28 at.% composition is effective to strengthen the composite. The microstructures in **Fig. 4.15b** expose interfacial microsized pores.

The EDS elemental mapping confirms the phases present in the sintered alloy as shown in **Fig. 4.16**. The Ni and Al phases form the matrix where martensite plates are located. TiC are widely distributed isolated particles within the $Ni_3Al$ eutectic

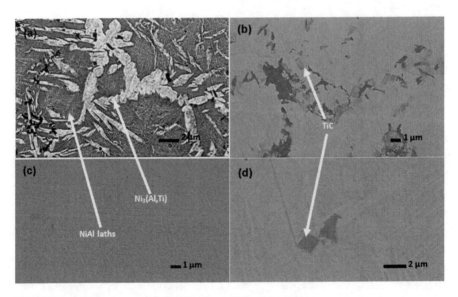

**FIG. 4.15**    SEM images of the 650 °C-sintered $Ni_{62.5}Al_{37.5}TiC_{1.28}$

phase with a small amount of Ti. Si contamination is due to the contamination during polishing.

$Ni_{62.5}Al_{37.5}$ is an Al-rich composition on the NiAl phase diagram as represented in **Fig. 4.17b**. Due to addition of TiC particles that impeded free movement in the microstructure, martensite structure formed was evident in **Fig. 4.17a**. The $Ni_{62.5}Al_{37.5}TiC_{1.28}$ composite resisting the cutting by the diamond-cutting tool is an indication of the work hardening. This behaviour is like those shown by the Fe-Mn steels strengthening mechanism [21] due to intragranular brittle $Ni_3Al$ and TiC precipitates [23, 24]. The martensite phase has developed due to thermally activated dislocation motion by the movement of atoms over distances that are less than the inter-atomic spacing [25]. Mostly, the NiAl martensite laths are exposed after water quenching (WQ) due to residual stress [26]. The WQ promotes the $Ni_3Al$ precipitation at grain boundaries. Cui et al. [27] reported the twinning and multiple dislocations dictating the strain hardening mechanisms in austenite, which in the current study was induced by TiC nanoparticles during the TE. The TiC particles and the $Ni_3(Al, Ti)$ phase have enclosed the martensite laths restricting further deformation, hence it is difficult to cut the material. The precipitation of the faceted TiC and $Ni_3(Al, Ti)$ particles is in agreement with literature [28].

**Fig. 4.18** shows the microhardness of the carbide free $Ni_{62.5}Al_{37.5}$ and $Ni_{62.5}Al_{37.5}TiC_{1.28}$ composite. Addition of 1.28 at.% increased the hardness of the composite drastically. All samples were sintered at 650 °C for 4 h. The $Ni_{62.5}Al_{37.5}TiC_{1.28}$ composite has an average hardness of 431 HV while the $Ni_{62.5}Al_{37.5}TiC$ is 396 HV. The hard $Al_3Ni$ intermetallic phase is not brittle when compared to carbides. The

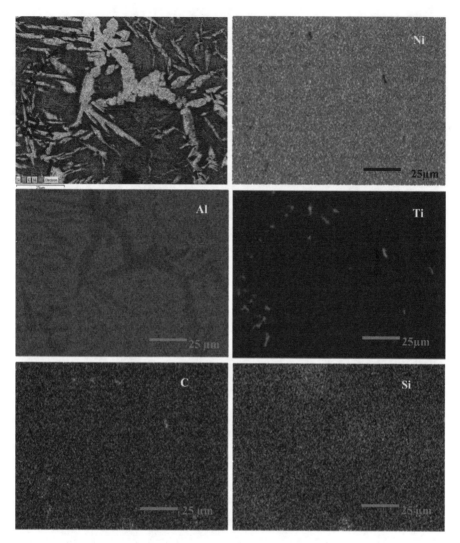

**FIG. 4.16** SEM-EDS elemental mapping of the 650 °C-sintered $Ni_{62.5}Al_{37.5}TiC_{1.28}$ composite

Hall-Petch relationship shows that the hardness of the composite is dependent on the grain size (Eq. 4.5), chemical composition, and interfacial pores present at interfaes [29]. The current hardness values were compared with literature in **Table 4.3**. The $Ni_3Al$ phase is harder than the NiAl phase. It is evident that the hardness of the TiC reinforced NiAl composite depends on the composition of the matrix as well as the TiC content [30–32].

$$Hv = H_O + kD^{-1/2} \qquad (4.5)$$

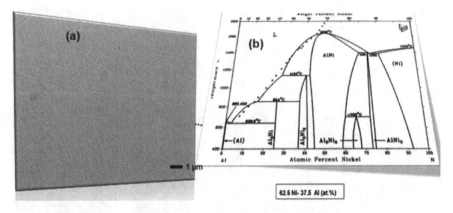

**FIG. 4.17**   Formation mechanism of the Ni$_{62.5}$Al$_{37.5}$TiC composite

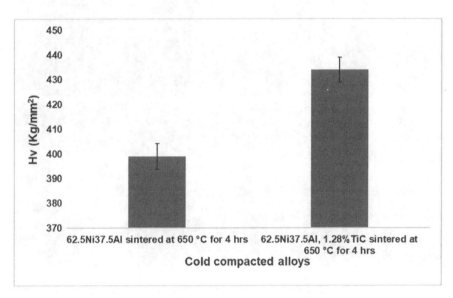

**FIG. 4.18**   Microhardness of the carbide free Ni$_{62.5}$Al$_{37.5}$ and Ni$_{62.5}$Al$_{37.5}$TiC$_{1.28}$ composite

## 4.4  BALL MILLING, MECHANICAL ALLOYING, AND POWDER ROLLING

### 4.4.1  BALL MILLING

Ball milling (BM) is a process of grinding lump materials in millimetres into micron or nano-size particles. As shown in **Fig. 4.19**, the process comprises the milling jar, grinding balls, and powder particles. The effectiveness of grinding depends on a number of factors: grinding speed, ball-to-powder ratio (BPR), and the milling

**TABLE 4.3**

**The hardness of the $Ni_{62.5}Al_{37.5}TiC_{1.28}$ composite compared with literature data**

| Alloy | Process | Hardness (HV) | Ref |
|---|---|---|---|
| NiAl | BM-vacuum hot pressing | 360±19 | [10] |
| TiB2-NiAl | Vacuum arc melting | 670 | [30] |
| | As-cast | 55 | |
| | Heat-treated | 361 | |
| NiAl | Cold isostatic pressing | 244.7 | [31] |
| NiAl-86 vol.%TiC | | 1428 | [31] |
| Ni–25 at.%Al (Ni3Al) | Vacuum arc melting furnace | 307±10 HV0.1 | [32] |
| Ni–25 at.%Al–7 wt.%TiC | Vacuum arc melting furnace | 449±10 | [33] |
| Ni–25 at.%Al–10 wt.%TiC | Vacuum arc melting furnace | 435±12 | [33] |
| NiAl | | 320 ± 19 | [8] |

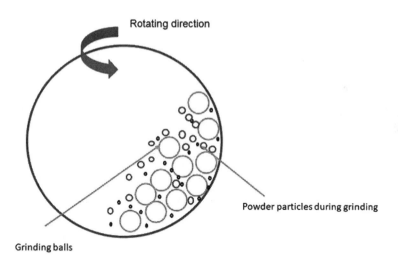

**FIG. 4.19** Schematic illustration of ball milling operation

atmosphere. In a large scale, the process is used in the mining industry to break ore into micron-size particles to enable easier physical or chemical extraction of pure metals. This process is well defined, and the mechanisms are described in literature. However, especially when done at high milling speeds, high-impact BM induces alloying between powder particles.

This behaviour is very convincing when interstitial elements like carbon (C), nitrogen (N), oxygen (O), and hydrogen (H) are added to the mixture. When tungsten (W) 99.99% and carbon (C) (4.2, 6.13, 17, and 23 wt.%) powders were milled at 250 and 300 revolutions per minute (rpm) in an argon atmosphere, the change in the material's crystal structure was validated by XRD analysis. A stoichiometric W-C composition was adopted. In 4 h of BM, the WC XRD peaks had emerged.

The Bragg peaks of W decreased as the C reacted with W while the WC formation completed after 8 h. If the milling jar is opened immediately after BM, hot powder is observed due to the exothermic reaction of WC. Interestingly, milling for 4 h, W peaks are still visible. The presence of unreacted W after 4 h shows the inhomogeneity of powders in early stages of milling because of high energy BM. At complete WC synthesis, the $WC_{0.83}$ phase was also found since there is also unreacted C in the mixture. The $WC_{0.83}$ compound has $B_h$ crystal structure with bigger lattice constants due to the higher amount of W than in WC [34]. The $WC_{0.83}$ peaks vanished after BM for 12 h but only the WC compound was detected [34].

The Scherrer equation was used to estimate the powder crystallite size. The XRD peak broadening was calculated from the full width at half maximum (FWHM) of the most intense Bragg peak. The Scherrer formula used is:

$$D = 0.9\lambda/B \, Cos\theta \tag{4.6}$$

where: $\theta$ = Diffraction angle, D = Crystallite size, $\lambda$ = X-ray wavelength, B = FWHM.

### 4.4.2 MECHANICAL MILLING (MM) OF ELEMENTAL POWDERS

MM refines the particle sizes [35] and induces the phase transformation [36–40] in elemental powders. In recent years, nanocrystalline (nc, grains sized below 100 nm) metals have captured a great deal of attention due to their improved chemical, physical, and mechanical properties compared with that of ultrafine crystalline (ufc, grain sizes ranging from 100 to 500 nm) or microcrystalline (mc, grains sized above 500 nm) metals. For ufc and mc metals, strength is drastically enhanced when grain size is reduced, following the Hall-Petch relationship. Here, the pile-up of dislocations at grain boundaries is envisioned as the key mechanistic process underlying an enhanced resistance to plastic flow from grain refinement; when grain size is relatively large, greater stresses can be concentrated near adjacent grains due to multiple pile-up dislocations, leading to the decrease in yield stress. On the other hand, as grain size is further reduced into nc region, activities of lattice dislocation become less significant, providing yield stress deviating from the Hall-Petch relationship [41–43].

When similar BM parameters are used to do alloying of Ti and Ni, different results are obtained. In WC, the interstitial reaction happens, but in TiNi solid solution should occur to develop and alloy.

### 4.4.3 METAL–METAL BALL MILLING

Under equilibrium conditions, two metals are melted at high temperature to produce an alloy with new mechanical properties. In the case of Ti and Ni, their melting temperatures are 1668 and 1453 °C, respectively. For Ti50 at.%Ni composition, the melting temperature of 1310 °C with a possible eutectic temperature at 1118 °C occurs [44].

### 4.4.4 MECHANICAL ALLOYING (MA)

MA was developed around the 1960s by John Benjamin and co-workers at the Nickel Company (INCO) to make oxide-dispersed strengthening in nickel-based super alloys for gas turbines [45, 46]. Since then, MA has been used to synthesize and study phase transformation in several binary alloys [47–48]. Equilibrium and metastable phases have been reported in metal powders [49]. It has emerged as a processing technique for alloy powders, amorphous and nanocrystalline by milling. A schematic illustration of MA is shown in **Fig. 4.20**, similar to the one studied in [50].

MA is a process that enables powder deformation while at the same time promoting alloying in powder mixtures through fracturing and welding. The process involves high-speed ball impact in an enclosed system. The powder particles become flattened and elongated as well as refractured to produce nanocrystalline powder particles. Powder particles change shape due to plastic deformation as shown in **Fig. 4.21**. In this case atoms slip due to exerted force (high stress) induced by milling balls through breaking atomic bonds due to mobility of dislocations at a preferred direction within the grain. **Fig. 4.21** also illustrates cold pressing (CP) which often called cold is sintering. It is a deformation process using high forces and pressures to form a metal

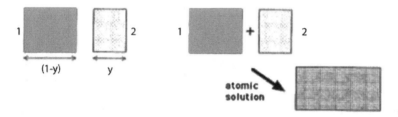

**FIG. 4.20** Formation mechanism of a solution after MA

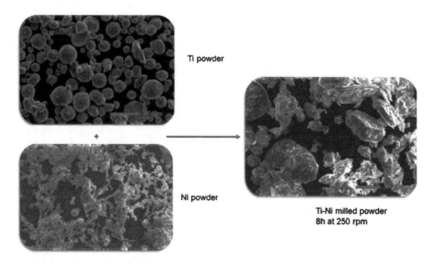

**FIG. 4.21** Ball-milled Ti plus Ni elemental powders

compound of a desired shape applied on powder particles in the unmilled or milled condition as illustrated in **Fig. 4.21.** For a mixture of pure elements with molar free energies $\mu_A$ and $\mu_B$, mechanical mixture is represented by equation 4.7, where x is the mole fraction of B.

$$G \ (mixture) = (1\text{-}x) \, \mu^\circ_A + x \, \mu^\circ_B \qquad (4.7)$$

### 4.4.5  BALL-MILLED TI POWDER

The BM process produces thermally reactive nanosized powder particles [51]. Therefore, handling of milled powders is one of the most important factors in prohibiting the particles' fresh surfaces to react with atmospheric air impurities. Pure Ti is reactive to interstitial oxygen (O), nitrogen (N), carbon (C), and hydrogen (H), hence, it is logical to attribute the HCP→FCC structural transformation to such impurities. Therefore, an extensive characterization after milling eliminates any doubt on the stabilization of metastable phases. The milling duration, milling speed, BPR, and milling environments (argon, air, and process control agents) influence the final crystal structure of the milled powder. For example, the low rotational speed (250 rpm) yields no phase transformation but nanocrystalline particles that are very reactive to thermal treatment. The nanocrystalline HCP Ti powder undergoes an HCP→FCC phase transformation on thermal treatment [52–54]. The metastable phases formed after BM are temporary structures that are sensitive to thermal treatment (annealing and sintering). Milled powders are characterized by the crystallite size and the crystal structure of the final powder depending on the severity of the milling process. The milled HCP Ti→FCC upon sintering is attributable to TiH$_x$ judged by the formation of needles [55]. An XRD lattice with parameters $a_{FCC} = 0.429$ nm and $a_{FCC} = 0.410$ nm was detected. The former FCC phase resembled the needle microstructure attributed to TiH$_x$ due to the presence of stearic acid. The latter was formed upon quenching and attributable to stresses introduced by rapid cooling. Therefore, two processes induced different Ti FCC lattice parameters. On the other hand, BM induces the HCP→FCC phase transformation. Seelam et al. [56] reported the HCP→FCC phase transformation in pure Ti powder during BM and ruled out that its stability is impurities-driven in two different experiments. One work was conducted with toluene to avoid agglomeration while the other powder was milled without toluene. The structural instability is due to negative hydrostatic pressure from grain refinement, increasing lattice expansion, and plastic strain/strain rate. Analysis on milled powder contamination done by the SEM-EDS technique at intermediate and final stages revealed no significant number of impurities [57]. The low-intensity P0 mill produced the HCP→FCC phase transformation after 150 h due to low energy milling at 250 h. High level of oxygen was detected in this powder due to adsorption of oxygen from the air during extended hours of milling [57]. Since then, many articles emerged in support of the HCP→FCC Ti phase transformation by milling. However, these reports differ in the lattice parameter of the FCC phase ~0.420 nm–0.440 nm. The positron annihilation spectroscopic tool validated the HCP→FCC Ti phase change was confirmed using the positron annihilation spectroscopic technique [57]. Despite these reports, the uncertainty on the FCC phase stability still prevailed until Seelam et al. [56] conducted two milling experiments under regular condition and under a high-purity argon atmosphere. Upon milling in air, the HCP→FCC phase transformation was obtained. However, milling under a high-purity argon atmosphere did not change the HCP phase leading to a conclusion that an FCC phase formed due to contamination [58]. The powder milled under normal air contained 9.4 at.%O, 6.4 at.%N and 2.3 at.%C. These contents are negligible when compared with equilibrium TiO compound (34.9–55.5 at.%O), TiN (28–50 at.%N) and TiC (~32–48.8 at.%C). The lattice parameter for milled Ti powder was a = 0.4238 nm, closely related to TiN, TiC, and $TiH_2$ with a = 0.4239 nm, a = 0.4327 nm, and a = 0.4431 nm, respectively. The lattice parameter is larger than that of TiO $a_{FCC}$ = 0.4177 nm [56]. The H content was not measured. The solid-state HCP→FCC phase transformation in Ti depends on the grain refinement and particle deformation like heating (thermal expansion) [59]. Also, the lattice parameter $a_{FCC}$ = 0.424 nm was achieved after BM but reverted to a rhombohedral structure upon sintering. It was proposed that the lattice parameter of FCC phase decreases with increasing milling time towards an ideal lattice constant of FCC Ti (a = 0.411 nm) as crystallite size reduces. This behaviour corresponds to reduced volume expansion and negative hydrostatic pressure as milling time increases [59]. The milling parameters during the BM of Ti powder are compared in **Table 4.4** across all the milling reports. On the other hand, Srinivasarao et al. [58] obtained a TiC FCC-like structure coexisting with α-Ti phase after BM that was stabilized upon hot pressing. The BM experiment induced high content of the FCC phase during milling. The TiC phase formed by BM and was confirmed by the significantly high C content [60–61]. A lattice parameter of a = 0.440 nm comparable to $TiH_2$ was obtained. The authors ruled out the possibility of $TiH_2$ formation due to the absence of an exothermic peak during DSC analysis. This justification further proved that thermal analysis conducted by Yazdani et al. [59] on milled HCP Ti represent the phase formation rather than HCP→BCC phase transformation. High number of impurities should suppress the HCP→FCC phase transformation during thermal analysis. This theory conflicts with research demonstrating that a Ti6Al4V alloy with H dissolved lowers its phase change temperature [62]. Therefore, the lower α→β phase transformation temperature in milled Ti25 at.%Al solid solution might be due to dissolved H. The ultrafine HCP Ti particles induced by BM are reactive upon annealing under nitrogen

**TABLE 4.4**

**List of the FCC lattice parameters induced by different process in pure Ti during milling**

| Material | Process | Crystal structure | Lattice parameter (nm) | Ref |
|---|---|---|---|---|
| Ti | BM (SA) annealed | FCC | 0.4291 | [55] |
| Ti powder | BM | FCC | 0.436 | [56] |
| Ti powder | BM | FCC | 0.433 | [57] |
| Ti powder | BM | FCC | 0.440 | [58] |
| Ti powder | BM | FCC | 0.424 | [61] |
| Ti powder | BM | FCC | 0.420 | [36] |

**TABLE 4.5**
**Types of milling equipment**

| Equipment | Container & Balls | BPR & PCA | Durations & Speed | Condition | Impurities | Refs |
|---|---|---|---|---|---|---|
| **SPEX 8000D** | SS | 10:1 (PCA) | 15 h | Air | EDS 6.4 N, 9.4 O, 2.3 C | **[58]** |
| **Horizontal simoloyer** | SS | 20:1(PCA) | 30 h-800 rpm | Ar | - | [59] |
| **attrition mill** | SS (AISI 304) | PCA | 5, 15, 30, 50, 75 h-450 rpm | Ar | (ICP-AES) Fe 1.15, 0.26 Cr, 0.18 Ni, 0.11 Mn | **[57]** |
| **Fritsch P6 planetary monomill** | WC | 10:1 with and without PCA | 30 h–250 h | Ar | EDS | [57] |
| **Horizontal simoloyer** | SS | 20:1 (PCA) | 15 h- 800 rpm | Ar | - | [55] |

SS = Stainless steel; WC = tungsten carbide; PCA = process control agent environment [62]. All work published under BM of Ti and Ti alloy powders shows that the milling parameters are not standardized (**Table 4.4**). For example, some authors used WC vials and balls, stainless steel balls, or hardened steel etc. The BPRs and milling speeds are also different while the analysis and measurement of impurities is not consistent. The determination of H content is very scarce.

**Table 4.5** and **Figs. 4.22–23** illustrate some of the equipment used for BM.

**FIG. 4.22**   Mechanical milling/alloying high energy ball mill machine

**FIG. 4.23**   Mechanical milling/alloying low energy ball mill machine

## Case study 2: Mechanically milled and sintered elemental Ti powder

*Introduction*

Ti possesses properties such as low density, MOE high biocompatibility, high corrosion resistance, and flexible mechanical properties. To achieve desirable mechanical properties, it is crucial to control the microstructure during processing. In its purity, Ti is favourable for use in dentistry devices and dental implant prosthesis components. This light metal shows better tissue tolerance than its competitors [63]. As early as the 1930s, Ti was used in biomedical application for dental implants [64, 65], due to its biocompatibility, corrosion resistance, and low MOE [65, 66]. However, its modulus is still considered high if compared to that of a human bone (30 GPa) [67]. This non-ferrous metal has a hexagonal closed-packed crystal structure (α-phase) which transforms into a body-centred cubic structure (β-phase) above 883 °C. This transformation has led to the innovation of the most successful Ti-6Al-4V in biomedical and aerospace application [67, 68]. Nevertheless, the high cost of Ti is still an industrial challenge. This high cost is associated with the high affinity of Ti to interstitial elements such as oxygen (O), nitrogen (N), carbon (C), and hydrogen (H) making it difficult to process [69]. As a result, scientific and technological interests to develop viable manufacturing methods are essential in cost reduction.

*Ball milling of pure Ti powder*

The commercially pure Ti powder with chemical composition is displayed in **Table 4.6.**

*Results and discussion*

**Fig. 4.24** shows the commercial Ti powder. The powder is smoothed out by gas atomization into a mixture of large and small spherical particles. After BM at 250 rpm speed for 16 h, these particles were broken and flattened, see **Fig.4.25**. This effect of milling is evident, indicative of deformation caused by impact of powder particles with balls and the wall of the milling jar. Detection of contamination by the EDS analysis was negligible.

## Characterization of Ti powder: Unmilled, cold-pressed, and sintered

The crystal structure of unmilled and cold-pressed pellet is shown **Fig. 4.26 (a)** and **(b)**, respectively. The XRD pattern detects only the HCP Ti ground-state crystal structure. The only difference between free and cold-pressed powder is the decreased peak intensity heights due to surface stresses in the cold-pressed sample.

**TABLE 4.6**
**Chemical analysis of pure titanium**

| N | O | C | Fe | Al | Cr | Ti |
|-------|-------|-------|-------|-------|------|---------|
| 0.003 | 0.009 | 0.003 | 0.025 | 0.008 | 0.01 | Balance |

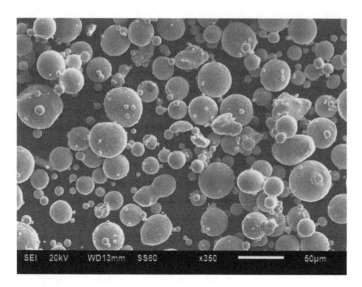

**FIG. 4.24** SEM image of the unmilled Ti powder

**FIG. 4.25** SEM image of the ball-milled Ti powder

The cold pressing can induce phase transformation [70, 71]. Cold pressing can induce alloying after repeated cold pressing as in the case of Sn-Te alloy [72]. The Ti powder's lattice parameters of a free and cold-pressed are a = 2.95; c = 4.68, and a = 2.92 Å; c = 4.67 Å, respectively. **Fig. 4.26 (c)–(d)** shows the XRD patterns of 900 and 1400 °C unmilled-sintered Ti samples, respectively. Only HCP crystal structure was detected.

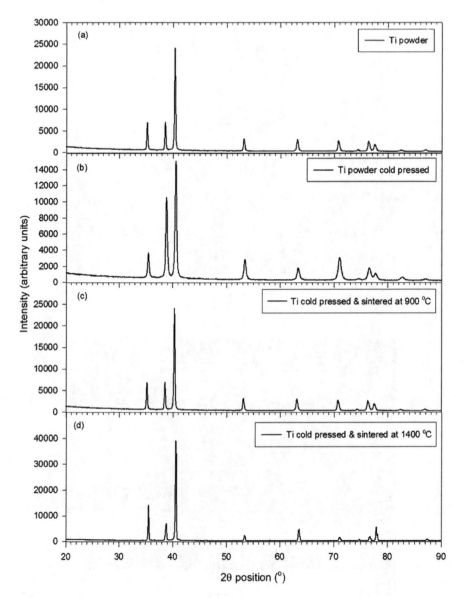

**FIG. 4.26**   XRD patterns of (a) unmilled, (b) cold-pressed, sintered at (c) 900 and (d) 1400 °C Ti, respectively

The XRD pattern of 900 and 1400 °C 16 h milled-sintered pellets at are shown in **Fig. 4.27(a)–(c)**, respectively. In addition to the HCP peaks, a new FCC phase has emerged. The calculated lattice parameter of the FCC phase is a = 4.292 Å. This lattice parameter falls in the range of FCC Ti induced by BM [**57, 73**]. However, this range also represents the FCC interstitial compound of Ti, both non-stoichiometry and stoichiometry is shown in **Table 4.7.**

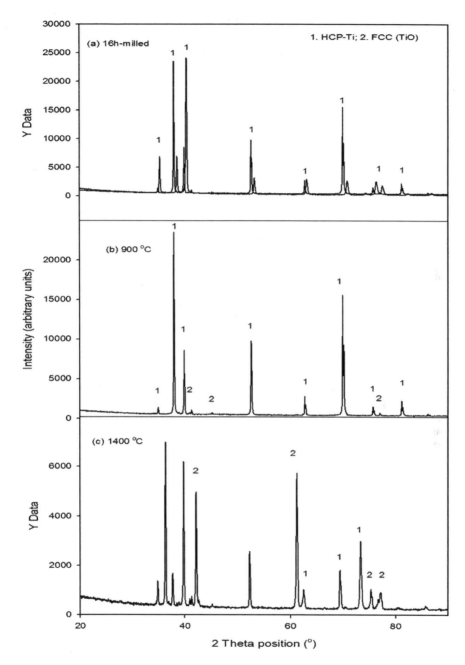

**FIG. 4.27** XRD patterns of (a) 16 h milled, sintered at (b) 900 and (c) 1400 °C Ti powder

**TABLE 4.7**
**Reported lattice parameters**

| Phase | Lattice parameter (Å) | Ref |
|---|---|---|
| **TiO** | 4.18–4.21 | [76] |
| **TiN** | 4.240 | [77,78] |
| **TiC** | 4.32 | [79, 81] |
| **TiH** | 4.10–4.44 | [73, 81] |
| **Ti (C, N)** | 4.22–4.35 | [82] |
| **Ti (O, N)** | 4.07–4.18 | [83] |

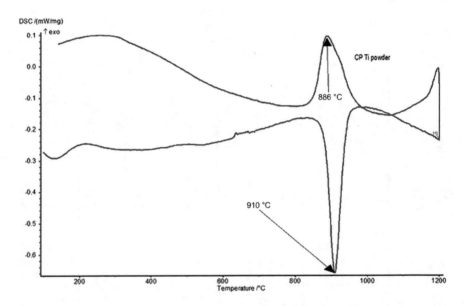

**FIG. 4.28**   DSC graph of Ti commercially CP unmilled loose powder

## Thermal analysis of Ti powders

Thermal analysis is a tool that indicates phase transformation behaviour and helps researchers to choose suitable sintering temperatures for materials. It is crucial to predict the sintering temperatures of both unmilled and milled powders. The DSC curve (**Fig. 4.28**) confirms the allotropic α→β phase transformation and indirectly ascertains the purity of the Ti. This transition shows an endothermic peak at 909 °C upon heating and an exothermic peak at 886 °C during cooling. This transformation validates reversible α→β to β→α transformation. The endothermic phase transition occurs upon heating [**19, 71–74**] similarly to second-order phase transitions [**19, 71, 72**]. This experiment was done on a free-flowing Ti powder; however, sintering is applied on pressed products. Therefore, the thermal analysis repeated on samples cut from the pressed pellets were conducted.

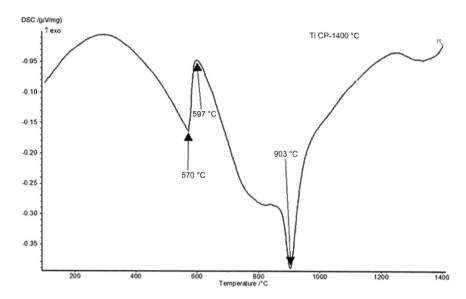

**FIG. 4.29** DSC graph of Ti cold-pressed powder

**Fig. 4.29** shows a non-linear DSC curve representative of a compacted pellet. A wide exothermic curve appears from the onset due to stress relieving. A sharp exothermic peak follows at 597 °C, alternating with α→β endothermic peak at 903 °C. It is occurring at a much lower temperature than that of free powder (909 °C). In the cold-pressed sample, the inter-particle distances are shorter than in free powder. Secondly, surface deformation due to cold pressing might have promoted this behaviour. Cold pressing occurs in a normal air flowing environment. Previous studies have indicated that surface deformation by means of external stresses on Ti promotes hydrogen dissolution [12, 73, 74]. On the 16 h milled Ti powder pellets, thermal analysis in **Fig. 4.29** shows the DSC with similar behaviour as in **Fig. 4.28**. External pressure by cold pressing changes the endothermic peak of the Ti powder.

Due to BM (external pressure), the phase transformation temperatures of α→β transition were reduced. The endothermic peak of the 16 h milled powder occurred at 692 °C. It implies that the β-transition shifted to lower temperatures. However, Vullum et al. [75] illustrated that excessive BM can inhibit the phase transformation. In **Fig. 4.30**, two overlapping endothermic peaks emerged between 1300 and 1400 °C attributable to partial melting. The milled-sintered compact shape changes due to partial melting.

### Microstructures of the sintered Ti samples
**Fig. 4.31** shows the optical microstructures of pure Ti sintered at 1400 °C. The optical microstructure reveals a columnar-type grain structure of different grain orientation.

The optical microstructure of the 16 h milled and sintered samples is shown in **Fig. 4.32**. Clear and precise grains defined by grain boundaries are evident. The grain

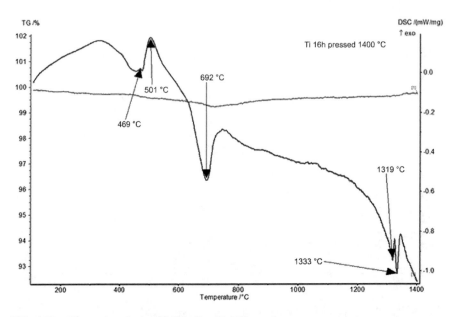

**FIG. 4.30**    Thermal analysis of 16h ball-milled Ti powder

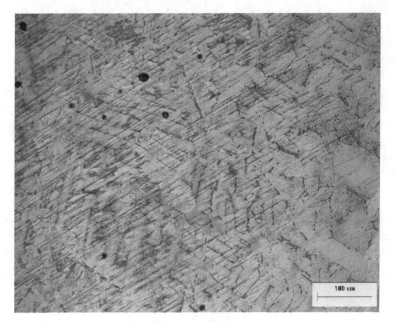

**FIG. 4.31**    Optical micrographs of a pure Ti sintered at 1400 °C

structure resembles the equiaxed-type grains. The second phase is segregated along the grain boundaries attributable to oxygen contamination.

The optical microstructure in **Fig. 4.33** displays the laths or plates. Laths in Ti are attributed to the possibility of hydrogen (H) contamination which the EDS technique

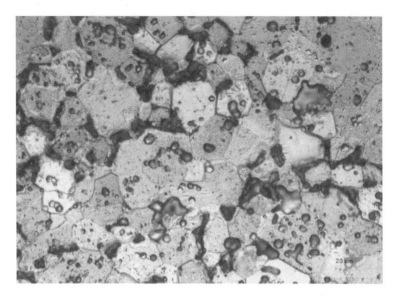

**FIG. 4.32** Optical micrographs of ball-milled 16h and sintered Ti at 900 °C

**FIG. 4.33** Optical micrographs of ball-milled 16h and sintered Ti at 1400 °C

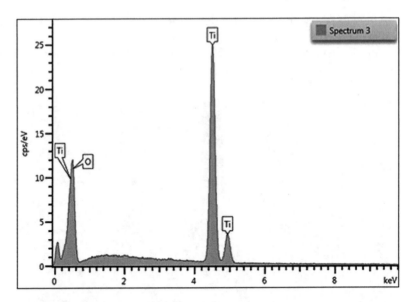

**FIG. 4.34**    EDS spectra done on 1400 °C milled and sintered Ti

is incapable of detecting. The EDS spectra of milled and sintered Ti at 1400 °C is shown in **Fig. 4.34**.

In addition to that, the lattice parameter of FCC $TiH_x$ depends on the H concentration; FCC phase can range from the smallest lattice [73] to the largest [74]. However, we assume that H and O dissolution have been picked from the stearic acid while carbon could have evaporated as CO. It seems H and O atoms were not detached from the $(CH_3(CH_2))_{16}COOH)$ chain by milling but could have dissolved during sintering. Due to insufficient amount of O to form, the stoichiometric compound, metastable FCC TiO contains vacancies that could be filled by hydrogen atoms. A TiH layer with an interface between HCP Ti and Ti (O, H) segregating the grain boundary was shown by Conforto et al. in [76]. The direct contact between Ti (O, H) and HCP Ti grains allows a different crystallographic relationship as shown on microstructures [28]. As a result of an increased sintering temperature (1400 °C), it is observable that grain orientation has changed from equiaxed to columnar. Hydride precipitation in HCP Ti produces volume change; hence, the misfit explains the needle shapes in optical microstructures [12]. The needles or plates become more distinct on an optical microscope than on the SEM [12]. The external pressure applied during MM could have unstabilized O and H in stearic acid, but dissociation occurs during heating. Luppo et al. [77], showed that 21 MPa external stresses can produce hydride reorientation. Upon thermal application, the presence of H and O on milled Ti favours the FCC crystal orientation resulting in metastable FCC Ti (O,H) with a lattice parameter of 4.29Å. The microstructure of 0 and 16 h milled Ti powder mixed at 1:1 ratio was analyzed. Similar studies have been reported in [78–82]. Suzuki and Nagumo showed that FCC TiC can be synthesized after reaction milling of titanium and n-Heptane $(CH_3(CH_2)_5CH_3)$ [83]. The effect of hydrogen was highlighted.

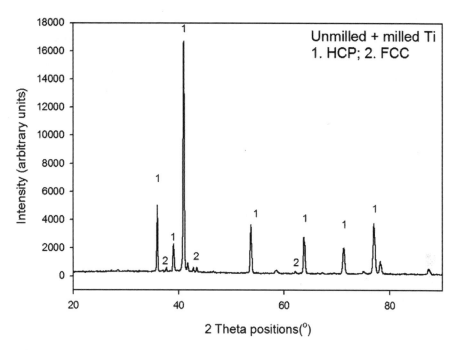

**FIG. 4.35**   XRD pattern of 1400 °C sintered (0) and 16 h Ti mixture

### Mixture of milled and unmilled powder

The XRD pattern of (0+16 h) Ti mixture sintered at 1400 °C is **Fig. 4.35**. Three phases, the HCP, FCC, and an unknown phase were detected. The HCP lattice parameters are smaller when compared to those of pure Ti phase. The obtained HCP lattice parameters were a = 2.885 Å; c = 4.624 Å. The milled powder particles are more reactive than the spherical unmilled as displayed by the DSC analysis. This was shown when milled Ti powder reacts with nitrogen gas at high temperature [51]. Furthermore, the thermal analysis shows a lower α→β phase transformation for the milled powder. The partial melting of milled powder particles upon sintering at 1400 °C may influence changes in crystal orientation during cooling when Ti undergoes reversion to an HCP structure, hence the reduced lattice constants. A small amount of FCC phase with lattice a = 4.22 Å was detected and attributed to TiO and detected by the EDS, **Fig. 4.36**.

**Fig. 4.37** shows the optical microstructure of the mixed unmilled and milled Ti powders. A bi-phasic optical microstructure with needle shaped particles has developed after sintering.

The macro-hardness of 16 h is higher than that of a pure and 0+16 h mixture Ti. An average of 654 HV on the polished surface was measured. The microstructure of the mixed sample has hardness values ranging between 500 and 600 HV.

### Metastable phases in elemental Co

A common metastable phase in elemental Co is an FCC crystal structure. This phase transformation is known to happen via thermal treatment as it can be illustrated by

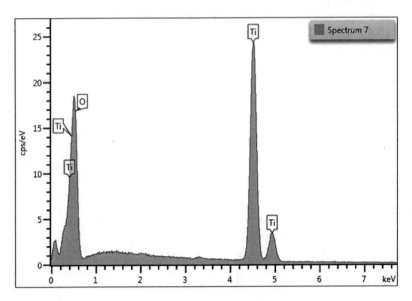

**FIG. 4.36**   EDS spectra of 0 and 16 h Ti mixture sintered at 1400 °C

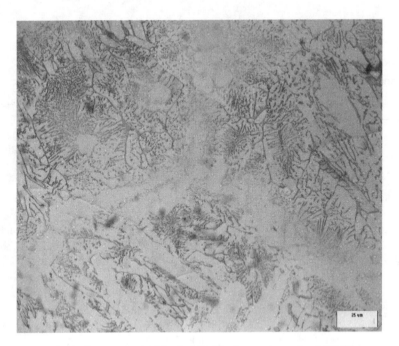

**FIG. 4.37**   Optical microstructure of the 0+16 h Ti mixture sintered at 1400 °C

the DSC technique [84]. Formation of this metastable phase at times co-exists with HCP depending on the grain size and the processing techniques [85, 86]. For the past decade, an effort to study pure metals by using non-equilibrium techniques has increased, especially MM. During milling, HCP Co transforms to a FCC phase, as reported by several researchers [11–13]. Nevertheless, thermal stability of mechanically milled Co is still outstanding in literature, hence the current project aiming to address that shortage.

## Metastable phases in W

A metastable beta (β) A15 W that serves as a superconductor can be synthesized via reduction by hydrogen and inert-gas condensation [87] and deposition [88]. Nanocrystalline and amorphous W have also been produced [89, 90]. The MM process has become a state-of-the-art technique recently used to synthesize amorphous and nanocrystalline materials that are commercially useful and scientifically interesting [91, 92]. Although nanocrystalline W powder has been produced via MM before, reports on thermal stability of milled powder are not available.

## Conclusions

The following conclusions were drawn:

- The milling effect is evident, indicative of deformation caused by impact of powder particles with balls and wall of the milling jar.
- The DSC curve confirmed the allotropic α→β phase transformation and indirectly ascertains the purity of the Ti, which validates the reversible transformation.
- The partial melting of milled powder particles upon sintering at 1400 °C influences changes in crystal orientation during cooling when Ti undergoes reversion to an HCP structure, hence the reduced lattice constants.
- Repeated cold pressing induces alloying as in the case of Sn-Te alloy.

## Case study 3: Annealing of powder mixture: Low pressure cold pressing

*Introduction*

*Al-Ti annealed powder* Lightweight structural materials used in the manufacture of aerospace, marine, and biomedical industry are important. In addition to light weight, high- and low-temperature strength, and wear and oxidation resistance are equally essential properties. Al-rich Ti [93] and Ti-rich Al intermetallic alloys [94] showed promising high-temperature properties. The Al-(Ti, Mg) is a catalytic agent and a diluent in the reactions [93]. Moreover, intermetallic phases crucial for high-temperature applications includes $Al_3Ni$, $Al_3Ti$, and $Al_2Ti$ [95] which are also additives for particle reinforced MMCs [96, 97]. The structural transformation and metastable phase formation in Al-rich Ti alloys produces the long-period ordered phases such as $Al_5Ti_3$, h-$Al_2Ti$, and r-$Al_2Ti$ in the L10 matrix upon annealing [98, 99]. For example, the lamellar structure composed of the L10 and h-$Al_2Ti$ in TiAl alloy forms due to annealing [98].

The $Al_5Ti_3$ metastable phase formed at 750–900 °C but dissolved above these temperature ranges [99]. $Al_2Ti$ is a stable intermetallic phase with a higher Al content. It has a lower density and excellent oxidation resistance when compared with $Al_3Ti$ phase [100]. The $Al_3Ti$ formed by the reaction and diffusion mechanisms during explosive welding, forming a strong bond between Al and Ti [95]. On the other hand, the $Al_3Ti$ structure is an additive intermetallic for Al/$Al_3Ti$ and Ti/$Al_3Ti$ composites to produce toughness and strength in the material [101]. Although the $Al_2Ti$ phase forms a lamellar structure, there is a metastable orthorhombic h-$Al_2Ti$ that exists below the tetragonal r-$Al_2Ti$ to h-$Al_2Ti$ transformation temperature [102].

The benefit of using PM in alloy development is to study the formation mechanisms and metastable phases in alloys. As a result, it is possible to identify a metastable phase in elementals as alloys either produced by MM or powder blending followed by annealing [103, 104]. The resultant intermetallic powders are potential additives for the development of Al MMCs [105, 106].

A mixture of A356 Al alloy, pure Ti (99.8%), Sn (99.7%), and Ni (99.95%) powders were mixed and compacted at 80 MPa. The alloy was composed of 94 wt.%Al, 4 wt.%Ti, 1 wt.%Ni and 1 wt.%Sn. The blended powder mixture was annealed/sintered at 700 °C in a tube furnace under flowing argon gas with 10 °C per minute heating rate. The annealed compacted pellets remained fragile upon sintering, and it was crushed into loose powder particles. The free powder particles were annealed at 400, 500, 600 and 700 °C for 6 h in air in a muffle furnace. The process of synthesizing the powders is displayed in **Fig. 4.38.**

A micromeritics AccuPyc II 1340 gas pycnometer was used to measure the true density of the sintered (700 °C) and annealed powders while the Microtrac Bluewave Particle Size Analyzer was used to measure the particle sizes. The oxygen (O) content of the powder was determined by means of ELTRA ONH 2000 PC under Helium gas atmosphere.

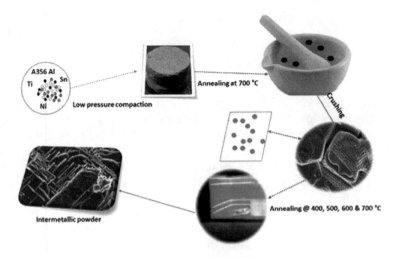

**FIG. 4.38**   Graphical representation of AlxTi powder production process

*Characterization of the annealed powder*

An SEM image of the Sn powder **Fig. 4.39** exhibits the mixture of spherical and irregular shapes particles. The A356 Al alloy powder reveal atomized spherical particles. The large spherical particles are cohabiting with ultrafine spherical particles and formed satellites. The Sn and A356 Al powder were mixed with spherical Ti and ultrafine powders **Fig. 4.40**.

The inset image in **Fig. 4.41b** shows the as-sintered fragile compact before crushing into particles as shown in **Fig. 4.41a**. Particle agglomeration is imminent indicative of the diffusion during sintering. As shown in **Fig. 4.41b–e**, crystalline and granular shaped particles have merged upon sintering. These particles appear as hacksaw-shaped platelets (**Fig.4.41b, c**). Moreover, different facet and platelet morphologies are evident in **Fig.4.41 d, e**. Notably, the A356 Al alloy powder is composed of ~7 wt.%Si and 0.35 wt.%Mg content. Despite the low melting temperature of Sn and Al alloy, there are possible immiscible reactions during alloying. The melting temperature of Sn and the A356 Al are 232 and ~557–613 °C, respectively. Upon sintering at 700 °C, eutectic reactions are activated leading to formations of $Al_xTi$ intermediate phases. This phenomenon is confirmed by the tetragonal HCP and BCC crystal structures detected by XRD analysis (**Fig. 4.42**).

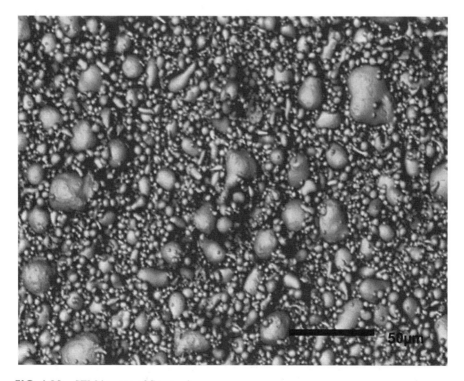

**FIG. 4.39** SEM images of Sn powder

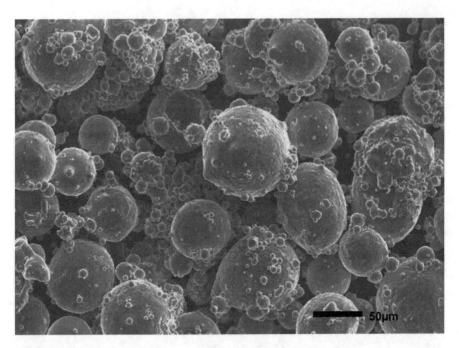

**FIG. 4.40**   SEM images of A356 Al powder

XRD analysis of the as-sintered powder has identified three phases as shown in **Fig. 4.42**. The dominating intensity peaks are of $Ti_5Si_3$ and $Al_3Ti$ intermetallic phases. The tetragonal $Al_3Ti$ intermetallic phase has lattice parameters $a = 3.848$ Å; $c = 8.596$ Å *(I4/mmm #139)* attributable to the $(AlSi)_3Ti$ phase [**105**]. The $Al_3Ti$ intermetallics are crucial for wear resistance in aluminium alloys. Their large grains are refined by adding strontium to improve the mechanical properties of the alloy [**107**]. Moreover, an HCP $Ti_5Si_3$ phase was identified with lattice parameters $a = 7.444$ Å; $c = 5.143$ Å *(P63/mcm #193)*. Minor BCC peaks with a lattice parameter $a = 2.868$ Å of *Im-3m # 229* $Fe_{0.94}Ni_{0.06}$ prototype was detected and attributed to BCC NiAl [**108**]. This NiAl phase forms at temperatures below 700 °C even when the alloy contains other alloying elements [**109**].

**Fig. 4.43a,b** shows the HR-TEM images revealing the presence of crystalline and amorphous grains. The presence of amorphous grains is linked to fine erratic XRD peaks. **Fig. 4.43b** shows facets and platelets that agree with the SEM analysis. The lattice fringes with d-spacing of 0.2395 nm corresponds to HCP $Ti_5Si_3$ (102) plane as detected by the XRD technique. In **Fig. 4.43a**, the SAED pattern (inset) contain both diffused and bright Debye rings, justifying the amorphous and polycrystalline phases. It also reveals a combination of single-crystal diffraction and polycrystalline in one image, justifying that one big crystallite might correspond to the sharp dots, with a small fraction of randomly oriented crystallites of the same material. The SAED pattern in **Fig. 4.43b** (inset) shows diffused Debye rings representing the amorphous structure.

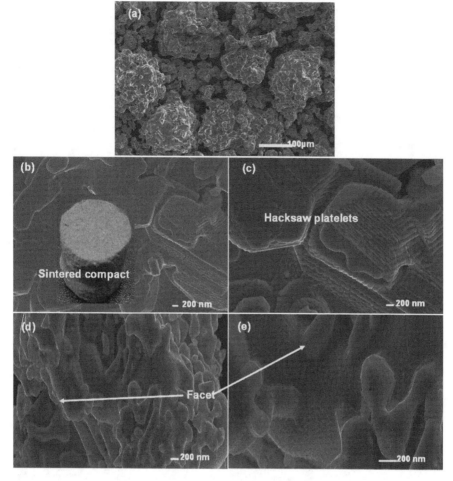

**FIG. 4.41**   SEM images of the annealed Al-Ti alloyed powder

To investigate the phase stability and oxidation resistance of the as-sintered powders, further heat treatment was conducted at 400, 500, 600, and 700 °C in air. The SEM images of the heat-treated powders are shown in **Fig. 4.44a-h**. The as-sintered hacksaw-thin platelet particles were changed into widely spaced laths/platelets upon annealing at 400 °C (**Fig. 4.44a, b**). The platelets are attributable to an $Al_3X$ phase. Moreover, irregular-shaped particles became agglomerated porous particles (**Fig.4.44a**). Upon annealing at 500 °C, the stepped thin platelets have emerged as shown in **Fig. 4.44c**. However, the irregular-shaped crystals appear to have different orientations (**Fig.4.44d**). Upon 600 °C (**Fig. 4.44e**), platelets were converted to curvy shapes while the spherical shapes mixed with porous particles has emerged as shown in **Fig. 4.44f**. The irregular-shaped particles formed large agglomerated grains with grain boundaries upon annealing at 700 °C (**Fig. 4.44g,h**).

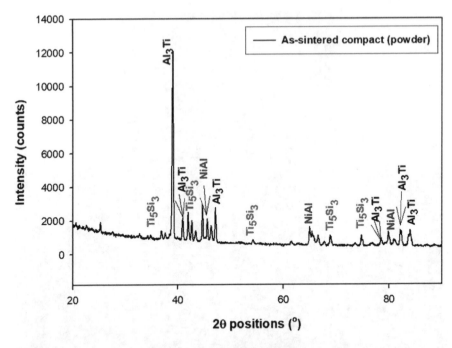

**FIG. 4.42**   XRD pattern of the alloyed powder upon sintering

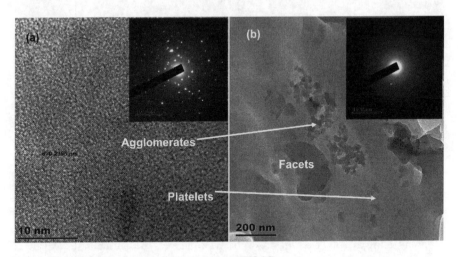

**FIG. 4.43**   HR-TEM of the as-sintered powder (700 °C)

The EDS analysis of the annealed powder confirmed the formation of major phases: r-Al$_2$Ti platelets and r-Al$_3$Ti (**Fig. 4.45a,b**). The carbon (C) was detected from the carbon tape used. The chemical composition of the r-Al$_2$Ti platelets is composed of an Al-rich phase with 65.70 ± 0.29 wt.%Al, 33.22 ± 0.14 wt.%Ti, 0.50 ± 0.21 wt.%Ni, and ± 0.58 ± 013 wt.%O. The composition of the r-Al$_3$Ti phase is 72.26 ±

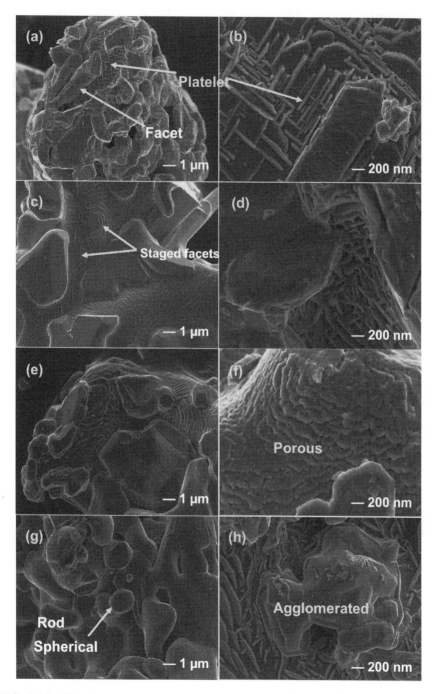

**FIG. 4.44** SEM images of the annealed powders at (a, b) 400, (c, d) 500, (e, f) 600, and (g, h) 700 °C in air

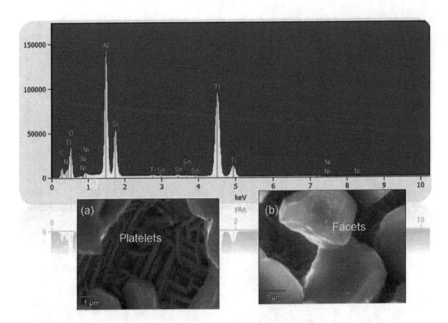

**FIG. 4.45**　EDS chemical line analysis of the annealed powders

0.15 wt.%Al, 21.24 ± 0.11 wt.%Ti, 3.95 ± 0.09 wt.%Ni, 0.36 ± 0.07 wt.%Sn ± 2.19 ± 1.13 wt.%O. The r-Al$_3$Ti phase dissolved more oxygen than the r-Al$_2$Ti phase. The N diffusion is attributable to the oxynitride layer formation [110]. Due to the presence of Si in the powder, the Ti$_5$Si$_3$ intermetallic phase has precipitated [111].

*Thermal stability of the powder annealed at 400, 500,600, and 700 °C in air*

**Fig. 4.46a–d** shows the XRD analysis of the 400, 500, 600, and 700 °C annealed powders. The 400 °C annealed powder (**Fig. 4.46a**) validates the EDS analysis. It confirms that phase transformation has occurred yielding tetragonal intermetallic r-Al$_3$Ti and r-Al$_2$Ti phases with lattice parameters $a$ = 3.848 Å; c = 8.596 Å and $a$ = 3.970 Å; c = 24.309 Å, respectively. All the lattice parameters are presented in **Table 4.8**. Additionally, an HCP phase with lattice parameters $a$ = 7.429 Å; $c$ = 5.139 Å attributed to Ti$_5$Si$_3$ formed. **Fig. 4.46b-d** shows the XRD patterns of the 500 °C, 600 °C and 700 °C powders. The r-Al$_2$Ti, r-Al$_3$Ti, and Ti$_5$Si$_3$ intermetallic phases remained stable on all the samples. The Ti$_5$Si$_3$ lattice parameter $a$ slightly changed with annealing temperature being 7.429 Å upon 400 °C and 700 °C, and 7.444 Å for 500 °C and 600 °C. In general, Al$_2$Ti, Al$_3$Ti and Ti$_5$Si$_3$ intermetallic phases were stable after annealing. It has been shown using the first-principles calculations that electronic and elastic properties of the r-Al$_2$Ti intermetallic has lattice parameters comparable with the current findings [112–114].

Despite the presence of low melting Sn in the powder mixture, affinity of Al to Ti favours the formation of Al$_2$Ti and Al$_3$Ti intermetallic phases. Sn acts as a catalytic agent and a diluent in the reaction [93]. The formation kinetics and diffusion of

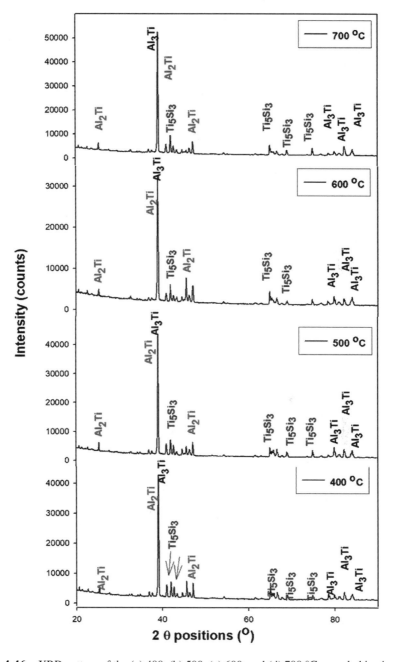

**FIG. 4.46**   XRD pattern of the (a) 400, (b) 500, (c) 600, and (d) 700 °C annealed in air

**TABLE 4.8**
**XRD data of the 700 °C sintered powder and of those powder after annealing at 400, 500, 600, and 700 °C**

| Sample ID | Phase | A (Å) | b(Å) | c(Å) | PDF reference code |
|---|---|---|---|---|---|
| **Sintered 700 °C** | $Al_3Ti$ | 3.848 | - | 8.596 | 03-065-2667 |
| **(powder)** | $Ti_5Si_3$ | 7.444 | - | 5.143 | 00-029-1362 |
| | BCC | 2.868 | - | - | 00-037-0474 |
| **400 °C** | $Al_3Ti$ | 3.848 | - | 8.596 | 03-065-2667 |
| | $Al_2Ti$ | 3.970 | - | 24.309 | 00-052-0861 |
| | $Ti_5Si_3$ | 7.429 | - | 5.139 | 00-008-0041 |
| **500 °C** | $Al_3Ti$ | 3.848 | - | 8.596 | 03-065-2667 |
| | $Al_2Ti$ | 3.970 | - | 24.309 | 00-052-0861 |
| | $Ti_5Si_3$ | 7.444 | - | 5.1430 | 00-029-1362 |
| **600 °C** | $Al_3Ti$ | 3.848 | - | 8.596 | 03-065-2667 |
| | $Al_2Ti$ | 3.970 | - | 24.309 | 00-052-0861 |
| | $Ti_5Si_3$ | 7.444 | - | 5.1430 | 00-029-1362 |
| **700 °C** | $Al_3Ti$ | 3.848 | - | 8.596 | 03-065-2667 |
| | $Al_2Ti$ | 3.970 | - | 24.309 | 00-052-0861 |
| | $Al_5Si_3$ | 7.429 | - | 5.139 | 00-008-0041 |

$Al_3Ti$ is faster than that of NiAlTi and $Ti_5Si_3$ phase formations. Due to the number of possible immiscible combinations in the current alloy system, we predict the less rigid compact formation which was easy to crush into powder. The heat treatment of the sintered powders conducted at 400, 500, 600, and 700 °C led to the formation of the BCC NiAl and tetragonal r-$Al_2Ti$ phases. Due to the lower free energy formation of the r-$Al_2Ti$ intermetallic phase, it has formed upon heat treatment but not after sintering at temperatures as low as 400 °C. This phase occurred through solid-state reactions of Al, Ti, Si, Ni, and Sn. This phase was reported stable at ~ 58 to 62.5 at.%Al composition in the temperature range between 700 °C and 1200 °C [97]. The immiscibility of Sn-Al and Sn-Si combinations provided a smooth reaction path for $Al_3Ti$ intermetallic phase formation during sintering [115, 116]. Therefore, segregation and formation of metastable phases were likely to form [117]. Even though Mg can evaporate out of the A356 alloy upon melting, it is also immiscible with Ti [118]. As a result, a metastable BCC AlNiTi phase was formed and undergoes further phase transformation upon annealing. It has been shown that cold pressing forces the softer powder particles with low melting temperatures to migrate to the compact surfaces [108]. The low melting temperature elements has low yield strength hence easily deformable [108].

The as-sintered powder has an average density measured at 3.035± 0.032 g/cm³ which upon annealing (heat treatment) at 400 °C has slightly increased to 3.042± 0.011g/cm³. The 500, 600, and 700 °C heat-treated powders' measured densities were 3.358±0.043 g/cm³ 3.285 ± 0.023 g/cm³, and 3.258 ± 0.017 g/cm³, respectively (**Table 4.9**). The density values are slightly different due to intermetallic concentrations. The density values are slightly lower when compared to the density of

**TABLE 4.9**
**Density measurements of the as-sintered alloyed and annealed powders**

| Sample ID | Density average | O content (wt.%) |
|---|---|---|
| **As-sintered at 700 °C** | 3.035± 0.032 | 0.199 ± 0.002 |
| **Oxidized at 400 °C** | 3.042± 0.011 | 0.368 ± 0.021 |
| **Oxidized at 500 °C** | 3.358±0.043 | 0.750 ± 0.077 |
| **Oxidized at 600 °C** | 3.285±0.023 | 2.028 ± 0.489 |
| **Oxidized at 700 °C** | 3.258±0.017 | 2.374 ± 0.182 |

the reported $Al_2Ti$; 3.53–3.54 g/cmm$^3$ since our powder contained mixed intermetallic phases. In **Table 4.9**, the oxygen content of annealed powders validates oxidation on the powders. The higher the annealing temperature, the higher the oxygen content since more O was absorbed at high temperatures.

**Fig. 4.47a** shows the HR-TEM image of the 400 °C annealed powder. The particles reveal evidence of agglomerated tree-like facets and the platelets. These facets are widely distributed throughout the matrix in **Fig. 4.47b**. SAED images shown in the inset (**Fig. 4.47b,c**) indicates highly crystalline facets. The platelets are light in colour when compared to the facets with the thickness of ~ ±110 nm. The segregation of the dark facets are not uniformly distributed and agrees with the SEM images as shown in **Fig. 4.47b,c**.

*Conclusions*

The following conclusions were drawn:

- Sintering at 700 °C activates eutectic reactions leading to formations of $Al_xTi$ intermediate phases.
- HR-TEM images revealing the presence of crystalline and amorphous grains. The presence of amorphous grains is linked to fine erratic XRD peaks.
- SAED patterns contained both diffused and bright Debye rings, justifying the presence of amorphous and polycrystalline phases.

## 4.4.6 POWDER ROLLING

The PR process consist of a hopper which enables a continuous feed of metal powder to a pair of rolls [**46**]. The pair of rollers mechanically compresses the powder to a green body and a powder strip is produced. The metallurgical bonding occurs by pressing and shearing while particle deformation and welding becomes effective. The produced strip requires sintering to allow diffusion between atoms. For better mechanical and surface properties the strip undergoes re-rolling and is wound into a coil. The secondary processing may include carburization, infiltration, restamping, surface compression, and impregnation. The product of PR has common features to those of BM due to the deformation taking place and particle welding. XRD analysis of the rolled powder strip and of the BM powder will show the broadening of intensity

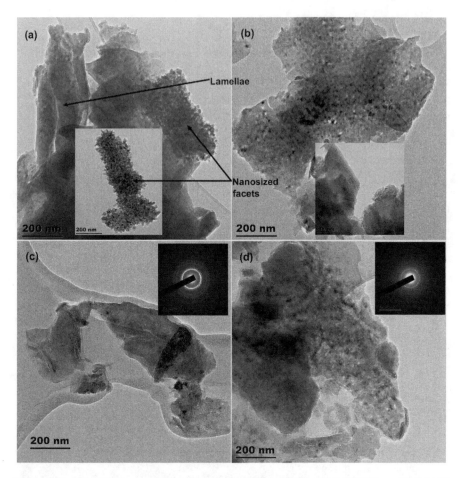

**FIG. 4.47**   HR-TEM images of the (a) 400, (b) 500, (c) 600, and (d) 700 °C annealed Al-Ti alloyed powder

peaks due to deformed particles. The schematic illustration of powder rolling is in Fig. 4.48. The rolled strip and the BM powder still require further sintering to achieve the mechanical properties desired for the material.

BM generates chaotic breaking by repeated stretching and folding operation of powder particles [119]. This indicates that MA is a very aggressive operation when compared to rolling [47]. Thus it is expected that metastable phases can be achieved after sintering of the rolled powder strip.

Parameters affecting powder rolling:

- Roll gap: Large roll gap leads to weak green density; very small roll gap leads to cracking of the strip on the edge
- Roll diameter: Increased density and strength with increase in roll diameter. for a given strip thickness

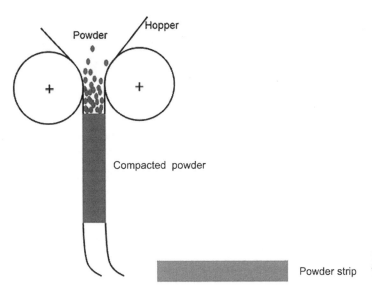

**FIG. 4.48**   Schematic drawing of a powder rolling process

**FIG. 4.49**   Powder rolling machine

- Roll speed: A balanced and tried speed is recommended for better strip properties
- Powder characteristics: Powder with rough surfaces provide better strip density

The direct powder rolling process can provide unique powder materials such as high-strength titanium alloys, aluminium alloys, and engine bearing alloys. The process has superior economic benefits with probability of technical success [**119**]. **Fig. 4.49** is a typical powder rolling equipment showing steel rollers.

## 4.5   METAL INJECTION MOULDING

In this process, the metal powder is mixed with binder material to create a feed-stock followed by being shaped and solidified using injection moulding of complex parts shaped in a single step. The product undergoes operations to remove the binder (debinding) followed by improving the density of powders. This process can undertake cost-effective mass production by producing complex near-net-shaped components from metal powders. The MIM has a combination of tooling design and manufacturing capabilities and capable of mass production of a wide range of small metallic parts with high dimensional accuracy. Schematic process illustration and instruments of the MIM process are shown in Fig. 4.50, and Fig. 4.**51**, respectively.

Fundamental requirements of MIM in each step of the process:

(1)  Metal powders
(2)  Binders
(3)  Mixing
(4)  Moulding
(5)  Debinding
(6)  Sintering

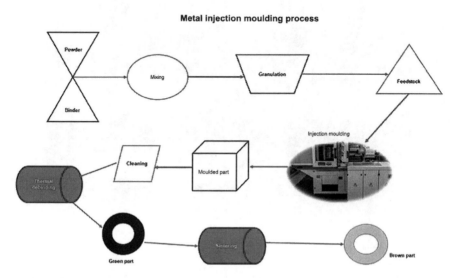

**FIG. 4.50**   Metal injection moulding process

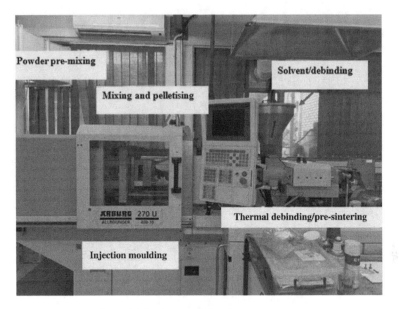

**FIG. 4.51** Illustration of metal injection moulding process

(7) Post-sintering operations
(8) Mechanical properties of MIM components

With injection moulding, however, the situation is quite different.

### 4.5.1 FEEDSTOCK PREPARATION

Before the MIM process starts, the suitable feedstock is prepared and mixed with fine metal powder, and blended with thermoplastic and wax binders in required amounts. The recommended 60:40 by volume metal powder to binder ratio is typically applied. The binders are melted, and the composition is mechanically mixed to uniformly coat the powder particles with the binders. This is followed by cooling and granulating into free-flowing pellets to be used in the MIM machine. In addition, capillary rheometry to examine the effects of temperature and shear rates on the flow nature of binder systems for a given metal MIM feedstock can be studied by the rheology equipment **Fig. 4.52**.

### 4.5.2 MOULDING

At high pressure, the pelletized feedstock is fed into an injection moulding machine where it is heated and injected into a mould cavity to produce a green part. The process is repeated to produce more green parts depending on the production rate. The mould cavity is designed approximately 20% larger to account for shrinkage that takes place during sintering, like patterns designed for metal casting.

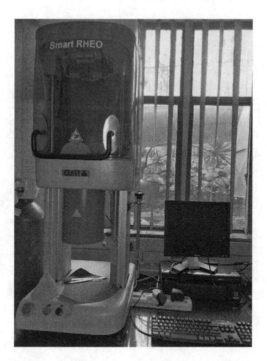

**FIG. 4.52**   Smart RHEO equipment

### 4.5.3   DEBINDING

The binder does not form part of the final product or the brown part, therefore, it must be removed, and the process is called debinding or the binder removal process. Binder removal is done in several steps before the sintering step, leaving behind only enough binder to handle the parts going into the sintering furnace. Therefore, some level of porosity is present after the binder removal. It is for this reason that only small parts can be produced which may not be used in high-strength applications.

### 4.5.4   SINTERING

During sintering, the parts, after debinding, are loaded into a high-temperature environment, inside the sintering furnace. The brown parts are heated gradually to remove the remaining binders before being heated to a high temperature to close the porosity by the diffusion process. Materials with a density close to theoretical densities can be produced. The fact that the powders used are very much finer in MIM than those used in PM means that sintering takes place more readily by reason of the higher surface energy of the particles. The brown part post debinding is very porous and this can result in a great deal of shrinkage. Therefore, the sintering temperature should be carefully controlled to retain the shape and prevent shrinking. The final product normally has a density closely approaching theoretical, greater than 95%. It is recommended that the parts produced are not used in structural applications where the material is exposed to extreme stresses

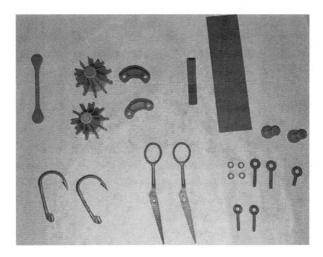

**FIG. 4.53**  Some artefacts produced by metal injection moulding

### 4.5.6  FINISHING

In normal circumstances, machining is not applied in the MIM process but may be employed depending on the final dimensions. Post-heat treatment processes are used to improve the properties of a materials. Moreover, plating, welding, and coatings can be successfully used. MIM is a competitive manufacturing process for small precision components which would be costly to produce by alternative methods. For example, it would be a tedious process to produce watch frames using gravity casting but rather convenient to use the MIM process. This includes complex shapes from a variety of materials such as ceramics, metals, composites, and intermetallic compounds. The practical application includes computer hardware, and biomedical, chemical, aerospace, and automotive equipment or machinery. **Fig. 4.53** illustrate some of the finished products.

## 4.6  CONCLUSIONS

PM is a process where powder particles are melded together to produce an alloy of specific and novel properties for an intended application. There are different types of powder processing with their different dynamics. These processes include press and sinter, mechanical alloying, metal injection moulding, powder rolling etc. Several factors dictate the success of the process to produce a desired product. All powder processes yield a part with green density which with further sintering at high temperature yields a metallic product with specific mechanical properties.

An alloy powder was produced by blending A356 Al, Ti, Sn, and Ni. The blend was compacted and sintered at 700 °C and crushed into powder. The sintered powder was annealed at 400, 500, 600, and 700 °C, respectively. The microstructure revealed r-$Al_2Ti$ facets and r-$Ti_3Al$ platelets phases. The phase transformation traced by the XRD identified the tetragonal $Al_3Ti$, HCP $Ti_5Si_3$, BCC NiAl-type phases. Minor peaks of the HCP $Ti_5Si_3$ phase were also detected. Upon annealing the powder in air, the tetragonal r-Al$_2$Ti showed better resistance to oxidation than the r-Ti$_3$Al phase. The emergence of the r-Al$_2$Ti after annealing resulted in widely spaced platelets in comparison to the stepped thin platelets. Annealing at 700 °C, the platelets became curved. The oxygen content of the powder increased with increasing annealing temperatures. These powders are a potential master alloy to be used in the production of a titanium and aluminium metal-matrix composite for high-temperature applications. Furthermore, thermal analysis conducted by the DSC revealed the lower α to β transformation temperature (692 °C) of mechanically milled titanium powder. After sintering at 900 °C, a microstructure with equiaxed grain orientation was observed. The grain orientation has changed to columnar after sintering at 1400 °C. XRD analysis detected the metastable FCC TiH phase which has segregated around the grain boundaries. The hybrid phase might have been promoted by the stearic acid. However, the EDS analysis could not detect any interstitial elements. On the other hand, the microstructure of the annealed Al-Ti powder revealed the r-Al$_3$Ti facets and platelets of the r-Al$_2$Ti phases. An alloy was produced by blending A356 Al, Ti, Sn, and Ni powders. The mixture was compacted, sintered at 700 °C and crushed into powder particles. Upon further annealing of the powder at 400, 500, 600, and 700 °C, r-Al$_3$Ti, r-Al$_2$Ti, and HCP Ti$_5$Si$_3$ intermetallic phases were formed. The Ni$_{62.5}$Al$_{37.5}$TiC$_{1.28}$ composite was developed by powder mixing, compaction, and sintering at 650 °C. The chemical reaction during sintering showed that thermal explosion occurred when small amount of nanosized TiC powder was added. The cold-pressing process reduced the activation/ignition temperature of the surface Ni/Al particle interaction, which propagated to the core of the compact. The resultant microstructure revealed martensitic NiAl plates surrounded by the hard Ni$_3$Al and TiC particles. The formed composite resisted the diamond tool cutting attributable to the work hardening behaviour of the composite. The Ni$_{62.5}$Al$_{37.5}$TiC$_{1.28}$ composite has an average hardness of 431±11 HV while the Ni$_{62.5}$Al$_{37.5}$TiC is 396±14 HV.

# REFERENCES

[1]  R. German, *Powder Metallurgy Science*, Princeton, New Jersey: Metal Powder Industry Federation, 451–455, 1994.

[2]  S. D. De la Torre, K. N. Ishihara, & P. H. Shingu, Synthesis of SnTe by repeated cold-pressing, *Mater Sci Eng A.*, 266, 37–43 1999.

[3]  A. Karin, A. Bonefacic, & D. Duzevic, Phase transformation in pressed cobalt powder, *J. Phys. F: Met. Phys.*, 14, 2781–2786, 1984.

[4]  C. Machio, M. N. Mathabathe, & A. S. Bolokang, A comparison of the microstructures, thermal and mechanical properties of pressed and sintered Ti-Cu, Ti-Ni and Ti-Cu-Ni alloys intended for dental applications, *Journal of Alloys and Compounds*, 848, 156494, 2020.

[5]  S. T. Camagu, A. S. Bolokang, T. F. G. Muller, D. E. Motaung, & C. J. Arendse, Surface characterization and formation mechanism of the ceramic TiO2-xNx spherical powder induced by annealing in air, *Powder Technology* 351, 229–237, 2019.

[6]  M. N. Mathabathe, G. Govender, C. W. Siyasiya, R. J. Mostert, & A. S. Bolokang, Surface characterization of the cyclically oxidized γ-Ti-48Al-2Nb-0.7Cr alloy after nitridation, *Materials Characterization*, 154, 94–102, 2019.

[7]   O. Ozdemir, S. Zeytin, & C. Bindal, Characterization of NiAl with cobalt produced by combustion synthesis, *J. Alloys Compd.*, 508, 216–221, 2010.

[8]   K. Bochenek & M. Basista, Advances in processing of NiAl intermetallic alloys and composites for high temperature aerospace applications, *Prog. Aero. Sci.*, 79, 136–146, 2015.

[9]   A. Abraham, H. Nie, M. Schoenitz, A. B. Vorozhtsov, M. Lerner, A. Pervikov, N. Rodkevich, & E. L. Dreizin, Bimetal Al–Ni nano-powders for energetic formulations Combust, *Flame*, 173, 179–186, 2016.

[10]  H. X. Dong, Y. H. He, J. Zou, N. P. Xu, B.Y. Huang, & C. T. Liu, Effect of preheating treatment at 575°C of green compacts on porous NiAl, *J. Alloy. Comp.*, 492, 219–225, 2010.

[11]  S. T. Camagu, N. M. Mathabathe, D. E. Motaung, T. F. G. Muller, C. J. Arendse, & A. S. Bolokang, Investigation into the thermal behaviour of the B2–NiAl intermetallic alloy produced by compaction and sintering of the elemental Ni and Al powders, *Vacuum*, 169, 108919, 2019.

[12]  A. S. Mukasyan, J. D. E. White, D. Kovalev, N. Kochetov, V. Ponomarev, & S. F. Son, Dynamics of phase transformation during thermal explosion in the Al–Ni system: Influence of mechanical activation, *Physica B*, 405, 778–784, 2010.

[13]  J. D. E. White, R. V. Reeves, S. F. Son, & A. S. Mukasyan, Thermal explosion in Al-Ni system: influence of mechanical activation, *J. Phys. Chem. A* 113, 13541–13547, 2009.

[14]  A. Aydogdu, Y. Aydogdu, & O. Adiguzel, Long-term ageing behaviour of martensite in shape memory Cu–Al–Ni alloys, *J. Mater. Proc. Technol.*, 153–154, 164–169, 2004.

[15]  A. Eftifeeva, E. Panchenko, Y. Chumlyakov, E. Yanushonite, G. Gerstein, & H. J. Maier, On the high cyclic stability of the tensile two-way shape memory effect in stress-induced martensite aged $Co_{35}Ni_{35}Al_{30}$ single crystals, *Mater. Sci. Eng. A.*, 799, 140166, 2021.

[16]  Yu Chao, Kang Guozheng, Rao Wei, & Song Di, Modelling the stress-induced multi-step martensite transformation of single crystal NiMnGa ferromagnetic shape memory alloys, *Mech. Mater.*, 134, 204–218, 2019.

[17]  A. S. Bolokang, M. N. Mathabathe, S. Chikosha, & D. E. Motaung, Investigating the heat resistant properties of the TiNi shape memory alloy on the B19′→B2 phase transformation using the alloy powder, *Surf. Interf.*, 20, 100608, 2020.

[18]  A. S. Bolokang & M. J. Phasha, Solid-state transformation in ball milled nickel powder, *Mater. Lett.*, 64, 1894–1897, 2010.

[19]  A. S. Bolokang & M. J. Phasha, Thermal analysis on the curie temperature of nanocrystalline Ni produced by ball milling, *Adv. Pow. Technol.*, 22, 2011, 518–521.

[20]  A. S. Bolokang, M. J. Phasha, S. T. Camagu, D. E. Motaung, & S. Bhero, Effect of thermal treatment on mechanically milled cobalt powder, *Int. J. Refract. Met. Hard Mater.*, 31, 258–262, 2012.

[21]  A. S. Bolokang, M. J. Phasha, D. E. Motaung, & S. Bhero, Effect of mechanical milling and cold pressing on co powder, *Journal of Metallurgy*, 2012, |Article ID 290873, | https://doi.org/10.1155/2012/290873.

[22]  R. Xiong, H. Peng, S. Wang, H. Si, & Y. Wen, Effect of stacking fault energy on work hardening behaviors in Fe–Mn–Si–C high manganese steels by varying silicon and carbon contents, *Mater. Des.*, 85, 707–714, 2015.

[23]  Y. Li, C. Li, J. Wu, Y. Wu, Z. Ma, L. Yu, H. Li, & Y. Liu, Formation of multiply twinned martensite plates in rapidly solidified $Ni_3Al$-based superalloys, *Mater. Lett.*, 250, 147–150, 2019.

[24] H. Zhao, F. Qiu, S. Jin, & Q. Jiang, High room-temperature plastic and work-hardening effect of the NiAl-matrix composites reinforced by particulates, *Intermetallics,* 19, (3) 376–381, 2011.

[25] H. L. Zhao, F. Qiu, S. B. Jin, & Q. C. Jiang, High work-hardening effect of the pure NiAl intermetallic compound fabricated by the combustion synthesis and hot pressing technique, *Mater. Lett.,* 65, 2604–2606, 2011.

[26] M. Clancy, M. J. Pomeroy, & C. Dickinson, Austenite and martensite microstructures in splat-cooled Ni-Al, *J. Alloys Compd.,* 523, 11–15, 2012.

[27] Y. X. Cui, L. Zhen, D. Z. Yang, G. P. Bi, & Q. Wang, Effect of quenching rate on microstructures of a NiAl alloy, *Mater. Lett.,* 48, 121–126, 2001.

[28] I. R. Souza Filho, M. J. R. Sandim, D. Ponge, H. R. Z. Sandim, & D. Raabe, Strain hardening mechanisms during cold rolling of a high-Mn steel: Interplay between submicron defects and microtexture, *Mater. Sci. Eng. A*, 754, 636–649, 2019.

[29] O. O. Marenych, A. G. Kostryzhev, Z. Pan, H. Li, & S. van Duin, Comparative effect of Mn/Ti solute atoms and $TiC/Ni_3(Al,Ti)$ nano-particles on work hardening behaviour in NiCu alloys fabricated by wire arc additive manufacturing, *Mater. Sci. Eng.,* 753, 262–275, 2019.

[30] B. M. Moshtaghioun, D. Gomez-Garcia, A. Dominguez-Rodriguez, & R. I. Todd, Grain size dependence of hardness and fracture toughness in pure near fully-dense boron carbide ceramics, *J. Eur. Ceram. Soc.,* 36 (7), 1829–1834, 2016.

[31] S. Talas, & G. Oruç, Characterization of TiC and TiB2 reinforced Nickel Aluminide (NiAl) based metal matrix composites cast by *in situ* vacuum suction arc melting, *Vacuum,* 172, 109066, 2020.

[32] M. X. Gao, Y. Pan, F. J. Oliveira, J. L. Baptista, & J. M. Vieira, Interpenetrating microstructure and fracture mechanism of NiAl/TiC composites by pressureless melt infiltration, *Mater. Lett.,* 58, 1761–1765, 2004.

[33] A. E. Karantzalis, A. Lekatou, & K. Tsirka, Solidification observations and sliding wear behavior of vacuum arc melting processed Ni–Al–TiC composites, *Mater. Charact.,* 69, 97–107, 2012.

[34] S. Bolokang, C. Banganayi, & M. Phasha, Effect of C and milling parameters on the synthesis of WC powders by mechanical alloying, *Int. Journal of Refractory Metals & Hard Materials*, 28, 211–216, 2010.

[35] E. Oda, H. Fujiwara, & K. Ameyama, Nano grain formation in tungsten by severe plastic deformation-mechanical milling process, *Materials Transactions,* 49, 54–57, 2008.

[36] P. Chatterjee & S. P. Sen Gupta, An X-ray diffraction study of nanocrystalline titanium prepared by high-energy vibrational ball milling, *Applied Surface Science,* 182, 372–376, 2001.

[37] I. Manna, P. P. Chattopadhyay, P. Nanadi, F. Banhart, & H. J. Fecht, Formation of face-centered-cubic titanium by mechanical attrition, *Journal of Applied Physics,* 93, 1520–1524, 2003.

[38] J. Sort, J. Nogues, S. Surinach, & M. D. Baro, Microstructural aspects of the hcp-fcc allotropic phase transformation induced by ball milling, *Philosophical Magazine,* 83, 439–455, 2003.

[39] J. Sort, J. Nogues, S. Surinach, J. S. Munoz, & M. D. Baro, Correlation between stacking fault formation, allotropic phase transformations and magnetic properties of ball-milled cobalt, *Materials Science and Engineering A,* 375–377, 869–873, 2004.

[40] J. Y. Huang, Y. K. Wu, H. Q. Ye, & K. Lu, Allotropic transformation of cobalt induced by ball milling, *Nanostructured Materials,* 6, 723–726, 1995.

[41]  V. Yamakov, D. Wolf, SR. Phillpot, & H. Gleiter, Grain-boundary diffusion creep in nanocrystalline palladium by molecular-dynamics simulation, *Acta Materialia.*, 50, 61–73, 2002.

[42]  Gleiter H. Nanostructured materials: state of the art and perspectives, *Nanostructured Materials,* 6, 3–14, 1995.

[43]  Gleiter H. Nanostructured materials: basic concepts and microstructure, *ActaMaterialia.*, 48, 1–29, 2000.

[44]  J. L. Murray, *Phase Diagrams of Binary Titanium Alloy*, Materials Park, Ohio: ASM International, 1987.

[45]  J. S. Benjamin, Fundamentals of Mechanical Alloying, *Materials Science Forum*, 88–90, 1–18, 1992.

[46]  P. S. Gilman & J. S. Benjamin, Mechanical Alloying, *Annual Review of Materials Science,* 13, 279–300, 1983.

[47]  E. Ma, Alloys created between immiscible elements, *Progress in Materials Science* 50, 413–509, 2005.

[48]  E. Ma & M. Atzmon, Phase transformation induced by mechanical alloying in binary systems, *Materials Chemistry and Physics,* 39, 249–267, 1995.

[49]  A. Y. Badmos & H. K. D. H. Bhadeshia. The evolution of solutions: A thermodynamic analysis of mechanical alloying, *Metallurgical Materials Transaction A,* 28, 2189–2193, 1997.

[50]  A. S. Bolokang & M. J. Phasha, Thermal analysis on the curie temperature of nanocrystalline Ni produced by ball milling, *Adv. Pow. Technol.* 22, 518–521, 2011.

[51]  A. S. Bolokang & M. J. Phasha, Formation of titanium nitride produced from nanocrystalline titanium powder under nitrogen atmosphere, *Int. J. Refract. Met. Hard. Mater.,* 28, 610–615, 2010.

[52]  V. V. Dabhade, T. R. Rama Mohan, & P. Ramakrishnan, Nanocrystalline titanium powders by high energy attrition milling, *Pow. Technol.*, 171, 177–183, 2007.

[53]  A. S. Bolokang, D. E. Motaung, C. J. Arendse, & T. F. G. Muller, Formation of the metastable FCC phase by ball milling and annealing of titanium–stearic acid powder, *Adv. Pow. Technol.*, 26, 632–639, 2015.

[54]  I. Manna, P. P. Chattopadhyay, P. Nandi, et al., Formation of face-centered-cubic titanium by mechanical attrition, *J. Appl. Phys.*, 93, 1520–1524, 2003. https://doi.org/10.1063/1.1530718.

[55]  I. Manna, P. P. Chattopadhyay, P. Nandi, & P. M. G. Nambissan, Positron lifetime studies of the hcp to fcc transformation induced by mechanical attrition of elemental titanium, *Phys. Lett. A,* 328, 246–254, 2004.

[56]  U. M. R. Seelam, G. Barkhordarian, & C. Suryanarayana, Is there a hexagonal-close-packed (hcp) face-centered-cubic (fcc) allotropic transformation in mechanically milled Group IVB elements?, *J. Mater. Res.*, 24 (11), 3454–3461, 2009.

[57]  M. J. Phasha, A. S. Bolokang, & P. E. Ngoepe, Solid-state transformation in nanocrystalline Ti induced by ball milling, *Mater. Lett.* 64, 1215–1218, 2010.

[58]  B. Srinivasarao, J. M. Torralba, M. A. Jabbari Taleghani, & M. T. Pérez-Prado, Very strong pure titanium by field assisted hot pressing of dual phase powders, *Mater. Lett.*, 123, 75–78, 2014.

[59]  M. B. Rahaei, R. Yazdani rad, A. Kazemzadeh, T. Ebadzadeh, Mechanochemical synthesis of nano TiC powder by mechanical milling of titanium and graphite powders, *Pow. Technol.*, 217, 369–376, 2012.

[60]  Y. Yang, H. Lu, C. Yu, & J. M. Chen, First-principles calculations of mechanical properties of TiC and TiN, *J. Alloys Compd.*, 458, 542–547, 2009.

[61] H. Liu, J. Cao, P. He, & J. C. Feng, Effect of hydrogen on diffusion bonding of commercially pure titanium and hydrogenated Ti6Al4V alloys, *Int. J. Hydro. Energy,* 34 (2), 1108–1113, 2009.

[62] J. C. Feng, H. Liu, P. He, & J. Cao, Effects of hydrogen on diffusion bonding of hydrogenated Ti6Al4V alloy containing 0.3 wt% hydrogen at fast heating rate, *Int. J. Hydro. Energy,* 32, 3054–3058, 2007.

[63] K. Wang, The use of titanium for medical applications in the USA, *Materials Science and Engineering A,* 213, 134–137, 1996.

[64] M. B. Nasab, M. R. Hassan, & B. B. Sahari, Metallic Biomaterials of Knee and Hip – A Review. *Trends Biomater. Artif. Organs,* 24, 69–82, 2010.

[65] R. Van Noort, Titanium: The implant material of today, *Journal of Materials Science,* 22, 3801–3811, 1987.

[66] M. Geetha, A. K. Singh, R. Asokamani, & A. K. Gogia. Ti based biomaterials, the ultimate choice for orthopaedic implants – A review. *Progress in Materials Science,* 54, 397, 2009.

[67] R. Van Noort, Titanium: the implant material today, *Journal of Material Science,* 22, 3801–3811, 1987.

[68] C. N. Elias, J. H. C. Lima , R. Valiev, & M. A. Meyers, Biomedical applications of titanium and its alloys, *Biological Materials Science JOM,* 60, 46–49, 2008.

[69] Y. Tomohiro, T. Threrujirapapong, I. Hisashi, & K. Katsuyoshi, Microstructural and mechanical properties of Ti composite reinforced with $TiO_2$ additive particles, *Transactions of JWRI,* 38, 37–41, 2009.

[70] P. Chatterjee & S. P. Sen Gupta, An X-ray diffraction study of strain localization and anisotropic dislocation contrast in nanocrystalline titanium, *Philosophical Magazine A,* 81, 49–60, 2001.

[71] D. Gu, W. Meiners, Y. Hagedorn, K. Wissenbach, & R. Poprawe, Structural evolution and formation mechanisms of TiC/Ti nanocomposites prepared by high-energy mechanical alloying, *Journal of Physics D: Applied Physics,* 43, 135402–135412, 2010.

[72] C. Papandrea & L. Battezzati. A study of the α to ϓ transformation in pure iron: rate variations revealed by means of thermal analysis, *Philosophical Magazine,* 87, (10), 1601, 2007.

[73] R. J. Contieri, M. Zanotello, & R. Caram, Recrystallization and grain growth in highly cold worked CP-Titanium, *Mater Sci Eng A,* 527, 3994, 2010.

[74] A. K. Rai, S. Raju, B. Jeyaganesh, E. Mohandas, R. Sudha, & V. J. Ganesan, Effect of heating and cooling rate on the kinetics of allotropic phase changes in uranium: A differential scanning calorimetry study, *Journal of Nuclear Materials,* 383, 215, 2009.

[75] P. E. Vullum, M. Pitt, J. Walmsley, B. Hauback, & R. Holmestad, Observations of nanoscopic, face centered cubic Ti and $TiH_x$, *Applied Physics A: Materials Science & Processing,* 94, 787–793, 2009.

[76] E. Conforto, D. Caillard, B.–O. Aronsson, & P. Descouts, Crystallographic properties and mechanical behavior of titanium hydride layers grown on titanium implants, *Philosophical Magazine,* 48, (7), 631–645, 2004.

[77] M. I. Luppo, A. Politi, & G. Vigna, Hydrides in α-Ti Characterization and effect of applied external stresses, *Acta Materialia,* 53, 4987–4996, 2005.

[78] G. A. Porter , P. K. Liaw, T. N. Tiegs, & K. H. Wu, Particle size reduction of the NiTi shape-memory alloy powders, *Scripta Materialia,* 43, 1111–1117, 2000.

[79] I. Veljković, D. Poleti, M. Zdujić, L. Karanović, & Ć. Jovalekić, Mechanochemical synthesis of nanocrystalline titanium monoxide, *Materials Letters,* 62, 2769–2771, 2008.

[80]  S. Piscanec, L. C. Ciacchi, E. Vesselli, G. Comelli, O. Sbaizero, S. Meriani, & A. De Vitta, Bioactivity of TiN-coated titanium implants, *Acta Materialia,* 52, 1237–1245, 2004.

[81]  S. Zhang, Titanium carbonitride-based cermets: processes and properties, *Materials Science and Engineering A,* 163, 141–148, 1993.

[82]  M. B. Rahaei, R. Yazdani Rad, A. Kazemzadeh, & T. Ebadzadeh, Mechanochemical synthesis of nano TiC powder by mechanical milling of titanium and graphite powders, *Powder Technology,* 217, 369–376, 2012.

[83]  T. S. Suzuki & M. Nagumo, Metastable intermediate phase formation at reaction milling of titanium and n-heptane, *Scripta Metallurgica et Materialia,* 32, (8), 1215–1220, 1995.

[84]  H. Matsumoto, Variation in transformation hysteresis in pure cobalt with transformation cycles. *Journal of Alloys and Compounds,* 223, 1–3, 1995.

[85]  E. A. Owen & D. Madoc Jones. Effect of grain size on the crystal structure of cobalt, *Process Physics Society B,* 67, 456–466, 1954.

[86]  S. Kajiwara, S. Ohno, K. Honma, & M. Uda, A new crystal structure of pure cobalt in ultrafine particles, *Philosophical Magazine Letters,* 55, 215–219, 1987.

[87]  W. Krauss & R. Birringer, Metastable phases synthesized by inert-gas-condensation, *NanoStructured Materials,* 9, 109–112, 1997.

[88]  T. Krabacak, P.-I. Wang, G.-C. Wang, & T.-M. Lu, Phase transformation of single crystal β-tungsten nanorods at elevated temperatures, *Thin Solid Films* 493, 293–296, 2005.

[89]  L. Magness, L. Kecskes, M. Chung, D. Kapoor, F. Biancianello, & S. Ridder, Behaviour and performance of amorphous and nanocrystalline metals in ballistic impacts, 19th International Symposium on Ballistics, 7–11 May 2001, Interlaken, Switzerland, 1183–1189.

[90]  K. Cho, L. Kecskes, R. Dowding, B. Schuster, Q. Wei, & R. Z. Valiev, Nanocrystalline and ultrafine grained tungsten for kinetic energy penetrator and warhead linear applications, Proceedings of the 25th Army Science Conference, Orlando, FL, 27 November 2006.

[91]  M. S. El-Eskandarany. *Mechanical alloying for fabrication of advanced engineering materials*, Norwich, New York: William Andrew Publishing, 2001.

[92]  R. W. Gardiner, P. S. Goodwin, S. B. Dodd, & B. W. Viney, Non-equilibrium synthesis of new materials, *Advanced Perfect Materials,* 3, 343, 1996.

[93]  Z. R. Yang, S. Q. Wang, X. H. Cui, Y. T. Zhao, M. J. Gao, & M. X. Wei, : Formation of Al3Ti/Mg composite by powder metallurgy of Mg–Al–Ti system, *Sci. Technol. Adv. Mater.,* 9, 1–6, 2008.

[94]  M. N. Mathabathe, A. S. Bolokang, G. Govender, R. J. Mostert, & C. W. Siyasiya, Structure-property orientation relationship of a $\gamma/\alpha_2/Ti_5Si_3$ in as-cast Ti-45Al-2Nb-0.7Cr-0.3Si intermetallic alloy, *J. Alloys Compd.,* 765, 690–699, 2018.

[95]  F. Foadian, M. Soltanieh, M. Adeli, & M. Etminanbakhsh, A study on the formation of intermetallics during the heat treatment of explosively welded Al-Ti multilayers, *Met. Mater. Trans. A.,* 45A, 1823–1832, 2014.

[96]  D. Himmler, P. Randelzhofer, & C. Körner, Formation kinetics and phase stability of in-situ $Al_3Ti$ particles in aluminium casting alloys with varying Si content, *Res. Mater.,* 7, 100103, 2020.

[97]  H. Chowdhury, H. Altenbach, M. Krüger, & K. Naumenko, Reviewing the class of Al-rich Ti-Al alloys: Modeling high temperature plastic anisotropy and asymmetry, *Adv. Mater. Modern Proc.,* 3, (16), 1–20, 2017.

[98]  T. Nakano, A. Negishi, K. Hayashi, & Y. Umakosh, Ordering process of $Al_5Ti_3$, h-$Al_2Ti$ and r-$Al_2Ti$ with FCC-based long-period superstructures in rapidly solidified Al-rich TiAl alloys, *Acta mater.*, 47, (4), 1091–1104, 1999.

[99]  F. Stein, L. C. Zhang, G. Sauthoff, & M. Palm, TEM and DTA study on the stability of Al5Ti3 and h-Al2Ti-superstructures in aluminium-rich TiAl, *Acta Mater.*, 49, 2919–2932, 2001.

[100]  J. E. Benci, J. C. Ma, & T. P. Feist. Evaluation of the intermetallic compound $Al_2Ti$ for elevated temperature applications, *Materials Science and Engineering A,* 192/193, 38–44, 1995.

[101]  L. C. Zhang, M. Palm, F. Stein, & G. Sauthoff, Formation of lamellar microstructures in Al-rich TiAl alloys between 900 and 1100 °C, *IntermetallicsI,* 9, 229–238, 2001.

[102]  Z. Wei, M. Yuan, X. Shen, F. Han, Y. Yao, L. Xin, & L. Yao, EBSD investigation on the interface microstructure evolution of Ti-$Al_3Ti$ laminated composites during the preparation process, *Mater. Charact.*, 165, 110374, 2020.

[103]  C. Mathebula, W. Matizamhuka, A. S. Bolokang, Effect of Nb content on phase transformation, microstructure of the sintered and heat-treated Ti (10–25) wt.% Nb alloys, *Int. J. Adv. Manuf. Technol*, 108, 23–24, 2020. https://doi.org/10.1007/s00 170-020-05385-9.

[104]  A. S. Bolokang & M. J. Phasha, Solid-state transformation in ball milled nickel powder, *Mater. Lett.*, 64, (17), 1894–1897, 2010.

[105]  S. Ma & X. Wang, Mechanical properties and fracture of in-situ $Al_3Ti$ particulate reinforced A356 composites, *Mater. Sci. Eng. A*, 754, 46–56, 2019.

[106]  Z. Gxowa-Penxa, P. Daswa, R. Modiba, M. N. Mathabathe, & A. S. Bolokang, Development and characterization of Al–Al3Ni–Sn metal matrix composite, *Materials Chemistry and Physics,* 259, 124027, 2021.

[107]  Y. P. Gong, S. M. Ma, H. J. Hei, Y. Ma, B. Zhou, Y. S. Wang, S. W. Yu, X. Wang, & Y. C. Wu, Tailoring microstructure and its effect on wear behavior of an Al–7Si alloy reinforced with in situ formed $Al_3Ti$ particulates, *J. Mater. Res. Technol.,* 9, 7136–7148, 2020.

[108]  S. T. Camagu, N. M. Mathabathe, D. E. Motaung, T. F. G. Muller, C. J. Arendse, & A. S. Bolokang, Investigation into the thermal behaviour of the B2–NiAl intermetallic alloy produced by compaction and sintering of the elemental Ni and Al powders, *Vacuum,* 169, 108919, 2019.

[109]   Z. Lu, F. Jiang, Y. Chang, Z. Niu, Z. Wang, & C. Guo, Multi-phase intermetallic mixture structure effect on the ductility of $Al_3Ti$ alloy, *Mater. Sci. Eng. A,* 721, 274–285, 2018.

[110]  A. S. Bolokang, D. E. Motaung, C. J. Arendse, & T. F. G. Muller, Production of titanium–tin alloy powder by ball milling: Formation of titanium–tin oxynitride composite powder produced by annealing in air, *J. Alloys Compd.* 622, 824–830, 2015.

[111]  M. N. Mathabathe, S. Govender, A. S. Bolokang, R. J. Mostert, & C.W. Siyasiya, Phase transformation and microstructural control of the α-solidifying γ-Ti-45Al-2Nb-0.7 Cr-0.3 Si intermetallic alloy, *J. alloys compd.*, 757, 8–15, 2018.

[112]  H. Sato & Y. Watanabe, Three-dimensional microstructural analysis of fragmentation behavior of platelet $Al_3Ti$ particles in Al-$Al_3Ti$ composite deformed by equal-channel angular pressing, *Mater. Charact.*, 144, 305–315, 2018.

[113]  J. Braun & M. Ellner, X-ray high-temperature in situ investigation of the aluminide $TiAl_2$ (HfGa$_2$ type), *J. Alloy Compd.*, 309, 118, 2000.

[114]  P.-Y. Tang, B.-Y. Tang, & X.-P. Su, First-principles studies of typical long-period superstructures $Al_5Ti_3$, h-$Al_2Ti$ and r-$Al_2Ti$ in Al-rich TiAl alloys, *Comp. Mater. Sci.*, 50, 1467–1476, 2011.

[115]  C. Eisenmenger-Sittner, H. Bangert, A. Bergauer, J. Brenner, H. Stfri, & P. B. Barna, *Vacuum,* 71, 253, 2003.

[116]   C. Eisenmenger-Sittner, H. Bangert, C. Tomastik, P. B. Barna, A. Kovacs, & F. Misjak, Solid state diffusion of Sn in polycrystalline Al films, *Thin Solid Films,* 433, 97, 2003.

[117]  Y. Nagae, M. Kurosawa, S. Shibayama, M. Araidai, M. Sakashita, O. Nakatsuka, K. Shiraishi, & S. Zaima, Density functional study for crystalline structures and electronic properties of $Si_{1-x}Sn_x$ binary alloys, *J. Appl. Phys.*, 55, 08PE04, 2016, https://doi.org/10.7567/jjap.55.08pe04.

[118]  C. Huang, H. Liu, R. Liu, T. Xi, B. Wu, Y. Mo, Z. Zhang, Z. Tian, & W. Liu, Simulation study of effects of Ti content on microstructure evolution and elastic constants of immiscible Mg-Ti alloys during rapid quenching process, *Mater. Lett.*, 220, 253–256, 2018.

[119]  P. H. Shingu (ed.) Special issue on mechanical alloying, *Mater Trans Japan Inst Metals,* 36, 83–388, 1995.

# 5 Advanced Machining Processes
## Machining Process Techniques

*A. S. Bolokang and M. N. Mathabathe*

## 5.1 INTRODUCTION

Advanced materials are developed to sustain performance improvement such as fuel efficiency in applications, for example, in the aerospace industry Yet, several of the advanced materials are under question when it comes to evaluating potential uses for aviation. Invariably, machining is a process of gradually removing the material from a workpiece, including metal cutting using either single or multi-point tools, with subsequent grinding with abrasive wheels comprised of a range of micro-cutting edges, randomly shaped and oriented [1].

## 5.2 CONVENTIONAL AND NON-CONVENTIONAL MACHINING METHODS

A variety of machining techniques experience unyielding tool wear as a result of elevated cutting force, incurring high machining costs and longer cutting times. Relative challenges of various methods [2] are shown in **Table 5.1**.

## 5.3 MACHINING PROCESS SELECTION AND MACHINABILITY

Today's manufacturing techniques have leaned towards a high mix of raw materials and depressed volume production, with resultant difficulty in process planning. Komatsu and Nakamoto [10] outlined that process planning includes the determination of cutting parameters, calculation of machining time and cost, selection of machine tools and cutting tools, and preparation of jigs and fixtures. The authors mentioned that machining characteristics are exhibited by using CAD models on both the workpiece and target shapes. This allows machining processes to ultimately enumerate the machining operations in the order of segmented machining [11].

The CNC milling machine surpasses other machines regarding surface quality and dimensional accuracy. However, good product quality depends on the combination of power consumption, material removal rate, tool life, surface roughness, and minimum

DOI: 10.1201/9781003356714-5

**TABLE 5.1**

**Feature comparison between conventional and non-conventional machining techniques**

| Type of machining technique | Material and/or information machined | Application | Conclusions | Reference |
|---|---|---|---|---|
| | | **Conventional** | | |
| Orthogonal cutting | • Carbon fiber-reinforced polymer/Ti alloy (CFRP/Ti-6Al-4V)<br>• Stack chip formation | Structural components in aircraft Industry or high tech applications | Increased cutting speed, and lowest depth of cut Reduced damaged CFRP phase with sound surface finish | [3] |
| Milling | • Ti-6Al-4V (grade 23) in beta annealed condition<br>• Super-critical CO2 as a coolant in face milling of the alloy | Medical industries | Tool life improved about 2.6 times the emulsion at the conditions tested. | [4] |
| Drilling | • Al2024-T3 fibre metal laminates<br>• Cutting tool coatings on surface roughness | Defence sectors | Wear mechanism seen in the drill flank was a combination of abrasion, coating delamination, and minor built-up edge | [5] |
| Grinding | • Hardened D2 tool steel<br>• Effect of grit size, feed rate, and depth of cutting | Tooling materials | Uniform wear distribution due to different mean-height of grains influencing the surface roughness | [6] |
| | | **Non-conventional** | | |
| Abrasive waterjet (AWJ) | • Ni-based superalloy<br>• Large number of embedded particles and scratches | Nuclear and marine industries | Localized plastic deformation, abrasive embedment modified the surface in terms of strengthening effects | [7] |
| Pulsed laser ablating (PLA) | • Al2O3 ceramics<br>• Influence of laser trepanning patterns on the hole ablating nano-second pulse laser | Information technology industries | Achieving high hole quality requires inlet diameters below 200μm or above 1000μm | [8] |
| Electrical-discharge machining (EDM) | • Al alloys viz. 2024-T351, 6061-T6, 5083-H116, and 7075-T651<br>• Effect of EDM on microstructure, and corrosion-fatigue | Corrosive environments | EDM surface acted as galvanic coupled to an adjacent bulk surface | [9] |

where $f$ is the cutting force, $ap$ is the cutting depth, and $v_c$ is the deflection.

**TABLE 5.2**
Some model correlations utilized for cutting mechanisms

| Machining approach | Theoretical model | Type of material | Process parameters | Remarks | Reference |
|---|---|---|---|---|---|
| PCBN tools with TiN coating | The ANN model | Ti-based superalloy | CS (m/min), FR (mm/rev), depth of cut (mm), SR (μm), TW (μm) | Minimum SR of 0.231μm is seen at 150mm/mm, 1mm/rev rate and 1mm, while the TW of 69.4426 is seen at 132m/min, 0.5mm/rev and 2mm | [16] |
| Femtosecond laser processing | Design expert software | Grooves on the SiC surface | Groove depth (μm), width (μm), HAZ (μm), and MRR (μm3/s) | The accuracy of process parameter prediction was validated due to the regression models used. | [17] |
| Abrasive water jet (AWJ) | 3D FE model | Carbon fiber-reinforced polymer (CFRP) | Orientation of the laminate, impact angle, and speed of water jet (m/s) | The findings can estimate the AWJ performance of laminates, and also be used for process control & optimize the parameters | [18] |
| CNC milling | 3D Solidworks modelling software (ANSYS static structural analysis) | Al alloy, Al-SiC graphite and Al-SiC composite materials | Cutting, feed and radial force, damping coefficient | Deformation and stress information is within specification and indicate longer life for Al-SiC composite machine bed | [19] |
| Edge intelligence brought by digital twin (DT) CNC machinery | Model partitioning and selection algorithm | Tool materials | Geometric parameters such as tool length and radius, spindle speed and feed rate, TW, and sensor data including acceleration, acoustic, and mechanical signals | The study establishes the construction and the use of DT model CNC systems | [20] |
| Abrasive water jet cutting | Multi-objective genetic algorithm | Carbon steel S235, 6.5mm thick | Vf = 127mm/min, ma = 300g/min, and h = 1mm, and t = 7.266s/cm2 | AWJ cutting performance was optimized | [21] |

where CS = cutting speed; FR = feed rate; SR= surface roughness; TW = tool wear; MRR = material removal rate; HAZ = heat affected zone; vf = transverse rate; ma = abrasive flow rate; h = standoff distance, and t = machining time required to produce a unit of cut surface.

time with limited resources [12]. A new four-stage femto-second laser drilling process optimization has been proposed by Wang et al. [13] whereby process efficiency and quality are concurrently enhanced, guided by machine learning with continuously changing parameters. **Table 5.2** lists some of the computer-based approaches employed for cutting mechanisms.

Effective prediction models of advanced materials have been established to generate the quantitative relationship between materials and machining processes, and enhanced strategies for high performance using modelling techniques were proposed [14]. For example, Taguchi grey relational analysis (TGRA) can be used on a variety of superalloys for machining through the abrasive water jet (AWJ) technique. Rana and Akhai [15] employed a multi-objective optimization of AWJ machining parameters for Inconel 625 alloy utilizing TGRA. The authors found that the best machining performance is attained with the largest mass flow rate, shortest standoff distance, and fastest transverse speed [15]. **Table 5.3** shows the machinability studies on some of advanced materials.

## 5.4   CHALLENGES RELATED TO MACHINING

Today's main challenges experienced by machining affect the establishment of advancing technological machining processes utilized by industry. Some of these challenges include tool wear and type of materials, such as ceramics which exhibit friable layers, surface cracks, and plastic deformation when processing by abrasive/grinding machining [29–30]. Dry machining is performed without the use of cutting fluids; though it has found favour recently, it exhibits challenges. These involve: (1) high temperatures at the tool-workpiece; (2) changes in dimensional accuracy; (3) catering for materials with excellent machinability; (4) hardness loss due to tool heating up; and (5) not being recommended for materials that do not yield acceptable results [31].

Though defined cutting edges using fibre-reinforced orientations of ceramic materials was investigated by Spriet [32], a negative outcome was experienced in this orthogonal cutting process. This included a ductile and brittle appearance of the ceramic machining showing grain fracture and slip plane mechanisms. Further, the coarseness of the machined surface led to micro-cracking as the scratch load was elevated [32]. C/SiC-based materials were machined by conventional milling, for a good surface finish and morphology. However, the results indicated the contrary, whereby, surface roughness increased at ~4µm penetration depth, culminating in brittle fracture due to larger groove formation [33]. To a greater extent, non-conventional machining invariably suffers challenges. For example, AWJ used for shaping and cutting hard materials was a useful technique in the work by Hashish et al. [34]. Despite that, the results highlighted that for effectiveness of this process for ceramic materials, the operating parameters have to be optimized, while PLA machining, usually for thermal damage reduction in hard materials, can be the best technique to use provided that the pulse durability, i.e. femtosecond laser ablations, is enhanced [35–36]. Lastly, for materials that are hard to cut, EDM comes in handy. However, process sparks are experienced due to temperature build-up deteriorating the machined surface, resulting in residual stress levels between the matrix and fibres [37].

**TABLE 5.3**
**A selection of machinability studies on advanced materials**

| Method | Material | Function | Type of cutting tool | Comments | Reference |
|---|---|---|---|---|---|
| Orthogonal cutting based CNC | Ni-based superalloy | Power generation applications | TiN coated cutting blade | Segmented chip formation was observed, but the distribution of particles in the Ni matrix resulted in short breaking chips | [22] |
| CNC lathe under wet conditions | Inconel 718 | Aircraft industries | Coated cemented carbide grade inserts | Tool life of the inserts reduced when feed speed was increased due to high cutting forces, abrasiveness, low thermal conductivity, and work hardening aspects | [23] |
| CNC CKA6150i lathe without lubricant and coolant | Ti-6Al-4V | Military and aerospace fields | Coated cemented carbide tool (CNMG120404-MA VP15TF) | Increasing cutting speed, the cutting temperature is affected by workpiece material softening rate and the bound workpiece on the tool surface | [24] |
| 2D finite-element model of hot machining | Ti-6Al-4V | Military components, automotive, and aerospace | Triangular shape cutting tool (TNMA 120404) fitted with MTGNR 2525M12 tool holder | Excellent correlation amid experiment and simulation is established. High process zone decreases flow stress of the material resulting in overall reduction of shear strength | [25] |
| Micro-mechanical machining | AISI 1005 & 1045 steel alloy | Engineering and structural materials | Flat-end mills (Fraisa M5712080) with a diameter of 800µm | The machinability of AISI 1045 was better compared to that of AISI 1005 steel, in terms of surface finish attributable to discontinuities during cutting and formation of grain boundary burrs. | [26] |
| Micro-electrical-discharge machining (M-EDM) | Zirconium boride reinforced with SiC fibres | Medical, dental prosthesis, & mechanical components | Solid WC electrodes with a diameter of 300µm were used as a tool | ZrB20 showed good surface roughness in compared to other materials due to low porosity resulting in improved mechanical properties | [27] |
| Conventional drilling process | Carbon fiber-reinforced plastic (CFRP) composites | Aerospace industry | WC drill with a cobalt binder | Machinability of AlCrN hard coating on drilling of CFRP has been investigated. The results showed that carbon fibers wears the coating unless parameters are adjusted. | [28] |

## 5.5 PRACTICAL ASPECTS AND GENERAL OUTLOOKS

Tool wear life-span in commercial metal cutting is an important aspect in sustaining good mechanical properties of advanced materials in aerospace applications. Recently, lubricant technology has been used to examine various tool wear lubricants, cutting force, and high-speed turning of any material. The cryogenic minimum quantity of lubricant (CMQL) technology was investigated by Zhang et al. [38], on a 300M steel. The outcome displayed a reduction of cutting force and temperature at the knifepoint and friction between the tool and workpiece, attributable to reduced tool wear and extended tool life. Furthermore, tool wear control at high-speed cutting 300M steel was examined using genetic algorithms-based methods. **Eq. 5.1** demonstrates a tool wear prediction:

$$VB = 0.0362v^{0.1054} f^{0.0567} a_p^{0.0119} \tag{5.1}$$

where $VB$ is the amount of flank water; $v$, $f$, and $a_p$ are the cutting speed, feed, and depth of cutting, respectively. However, if the influence of a factor is not significant on the dependent variable, then the coefficient of the factor should be zero, i.e., $bi = 0$, utilizing $t$ test as shown in **Eq. 5.2**:

$$ti = \frac{bi / \sqrt{cii}}{\sqrt{Q/(n-m-1)}}, i = 1, 2, \ldots\ldots, m \tag{5.2}$$

The tool wear coefficient level as ascertained in [38] was 0.05($\alpha = 0.05$) and t($\alpha$/2) (n-p-1), and the aggregate was t0.025(n-p-1) = 2.22814. This demonstrated the effect of the cutting speed on the forecast value, with subsequent feed and the depth of cutting. Lastly, to attain economic tool life with subsequent surface conditions, cutting parameters must be considered. For example, on powder metal Ni alloys [39], these are generally:

- Strain <0.01 mm
- Surface roughness <0.8 μm
- Non-parent material required
- No redeposited material or layer
- No light contrast amorphous or recast layer

It is apparent that machinability of micro-components is convenient if the fabricated parts are single-phase materials with low ductility comprised of fine grains in the microstructure. Albeit, in general, fabrication of micro-components from engineering materials embody multi-phase structures. For this shortfall, Mian et al. [26] suggested choosing materials having binary phases, one of which is predominated by a single phase and the other of which has a somewhat balanced volume percentage. As a result, the machinability of these materials provides insight into how phase characteristics affect the surface roughness and sub-surface microstructure during the micro-machining process [26].

**TABLE 5.4**
**Hybrid machining techniques**

| Dissimilar tool techniques/ energy sources | | Coordinated applications of process mechanisms | Reference |
|---|---|---|---|
| Reinforced processes e.g., | Assorted processes e.g., | Coordinated machining and rolling e.g., | [41–44] |
| Laser-reinforced water jet cutting | Laser cutting EDM | Mill-grind i.e., mechanical turning and milling | |
| Ultra-reinforced EDM | Abrasive-wire EDM (A-WEDM) | Turn-mill i.e., mechanical turning and milling | |
| Media-reinforced machining (MRM) | EDM/ECM combination | | |
| Vibration-reinforced machining (VRM) | HSM/ECM combination | | |
| Laser-reinforced machining (LRM) | | | |

Recently, hybrid processing integrates various manufacturing processes to fabricate artefacts efficiently and in a productive way. The driving force for the hybrid processes is resource and energy concerns, opening new prospects and applications for engineering components such as advanced manufacturing/composite systems, and gas turbines for the aerospace industry [1]. It is important to note that the integral processes in hybrid technology commune in the same machining zone concurrently. Some examples are listed in **Table 5.4.**

On-going efforts in structural health monitoring (SHM) systems with embedded sensors have been an advancement in engineering structures. However, there are some dependencies with adverse effects on the sensing properties including (1) base metal properties; (2) manufacturing processes used for integration; (3) framework structure in the midst of material and surface reinforced; (4) signal processing techniques; (5) mechanical and thermal loading [40]. In contrast, piezoelectric materials provide good durability resulting in excellent sensing mechanisms. Optical sensors, although costly, are systematic in measuring strain and temperature synchronously with subordinate top-ranking performance [40]. In addition, the multimaterial 3D emerging printing techniques alleviate the integration of sensor materials into 3D printed composites, henceforth promoting insight into the deformation of interfacial mechanisms of composite materials used in the architecture industry [40].

## 5.6   CONCLUSIONS

Advanced materials are developed to tolerate deleterious conditions such as environmental damage (creep, corrosion, and oxidation), dwell crack growth, creep strain, microstructure stability, and elevated temperature yield stress. These conditions are exclusive of escalating costs and density. On the other hand, the proposal for an improved machining method gained ground. For instance, additional improved mechanical properties, such as cost-adequacy, are required for Ni-based superalloys.

When it comes to tool production and application, issues with the tool failure mechanism during machining exist. Tool failure expedition can be subdivided into (1) macro problems of tool damage, and (2) the prior method used to create the surface crack defects and microcrack propagation into the substance. For example, failure mechanism of carbide tool during cutting process can be contemplated, namely, (1) crack initiation; (2) crack-propagation; (3) damage pile-up; and (4) tool breakage.

## REFERENCES

[1]   W. Grzesik, *Advanced machining processes of metallic materials.* Amsterdam, Netherlands: Elsevier B.V, 2017. doi: 10.1016/B978-0-444-63711-6.0000-8.

[2]   M. N. Mathabathe & A. S. Bolokang, Challenges in machining of advanced materials, *Adv. Sustain. Mach. Manuf. Process.*, 3–15, 2022, doi: 10.1201/9781003284574-2.

[3]   J. Xu, M. El Mansori, M. Chen, & F. Ren, Orthogonal cutting mechanisms of CFRP/Ti6Al4V stacks, *Int. J. Adv. Manuf. Technol.*, 103, (9–12), 3831–3851, 2019, doi: 10.1007/s00170-019-03734-x.

[4]   N. Tapoglou, C. Taylor, & C. Makris, Milling of aerospace alloys using supercritical CO2 assisted machining, *Procedia CIRP*, 101, 370–373, 2020, doi: 10.1016/j.procir.2020.06.008.

[5]   K. Giasin, G. Gorey, C. Byrne, J. Sinke, & E. Brousseau, Effect of machining parameters and cutting tool coating on hole quality in dry drilling of fibre metal laminates, *Compos. Struct.*, 212, (September 2018), 159–174, 2019, doi: 10.1016/j.compstruct.2019.01.023.

[6]   R. Hood, F. Medina Aguirre, L. Soriano Gonzalez, D. Novovic, & S. L. Soo, Evaluation of superabrasive grinding points for the machining of hardened steel, *CIRP Ann.*, 68, (1), 329–332, 2019, doi: 10.1016/j.cirp.2019.04.090.

[7]   Z. Liao, I. Sanchez, D. Xu, D. Axinte, G. Augustinavicius, & A. Wretland, Dual-processing by abrasive waterjet machining: A method for machining and surface modification of nickel-based superalloy, *J. Mater. Process. Technol.*, 285, (April), 116768, 2020, doi: 10.1016/j.jmatprotec.2020.116768.

[8]   W. Zhao & X. Mei, Optimization of trepanning patterns for holes ablated using nanosecond pulse laser in al2o3 ceramics substrate, *Materials (Basel).*, 14, (14), 2021, doi: 10.3390/ma14143834.

[9]   S. R. Arunachalam, S. E. Galyon Dorman, R. T. Buckley, N. A. Conrad, & S. A. Fawaz, Effect of electrical discharge machining on corrosion and corrosion fatigue behavior of aluminum alloys, *Int. J. Fatigue,* 111, (January), 44–53, 2018, doi: 10.1016/j.ijfatigue.2018.02.005.

[10]  W. Komatsu & K. Nakamoto, Machining process analysis for machine tool selection based on form-shaping motions, *Precis. Eng.*, 67, (September 2020), 199–211, 2021, doi: 10.1016/j.precisioneng.2020.09.023.

[11]  W. Komatsu, Y. Inoue, & K. Nakamoto, Proposal of computer aided process planning of parts machining using workholding devices, *Journal of Advanced Mechanical Design, Systems, and Manufacturing – Bulletin of the JSME*, 14, (3), 1–12, 2020.

[12]  S. Patil, P. Sudhakar Rao, M. S. Prabhudev, M. Y. Khan, & G. Anjaiah, Optimization of cutting parameters during CNC milling of EN24 steel with Tungsten carbide coated

inserts: A critical review, *Mater. Today Proc.*, 62, 3213–3220, 2022, doi: 10.1016/j.matpr.2022.04.217.

[13] C. Wang, Z. Zhang, X. Jing, Z. Yang, & W. Xu, Optimization of multistage femtosecond laser drilling process using machine learning coupled with molecular dynamics, *Opt. Laser Technol.*, 156, (June), 108442, 2022, doi: 10.1016/j.optlastec.2022.108442.

[14] Y. Chen, S. Wang, J. Xiong, G. Wu, J. Gao, Y. Wu, G. Ma, H. Wu, & X. Mao, Identifying facile material descriptors for Charpy impact toughness in low-alloy steel via machine learning, *J. Mater. Sci. Technol.*, 132, 213–222, 2023, doi: 10.1016/j.jmst.2022.05.051.

[15] M. Rana & S. Akhai, Multi-objective optimization of Abrasive water jet Machining parameters for Inconel 625 alloy using TGRA, *Mater. Today Proc.*, 65, (8), 3205–3210, 2022, doi: 10.1016/j.matpr.2022.05.374.

[16] V. Veeranaath, M. N. Mohanty, A. Kumar, & P. Kumar, ANN modeling of the significance of constraints in turning superalloys using coated PCBN tools, *Mater. Today Proc.*, 65, (1), 20–28, 2022, doi: 10.1016/j.matpr.2022.03.559.

[17] R. Zhang, C. Huang, J. Wang, D. Chu, D. Liu, & S. Feng, Experimental investigation and optimization of femtosecond laser processing parameters of silicon carbide–based on response surface methodology, *Ceram. Int.*, 48, (10), 14507–14517, 2022, doi: 10.1016/j.ceramint.2022.01.344.

[18] M. Demiral, F. Abbassi, T. Saracyakupoglu, & M. Habibi, Damage analysis of a CFRP cross-ply laminate subjected to abrasive water jet cutting, *Alexandria Eng. J.*, 61, (10), 7669–7684, 2022, doi: 10.1016/j.aej.2022.01.018.

[19] R. Kumar, A. Jain, S. K. Mishra, M. Joshi, K. Singh, & R. Jain, Comparative structural analysis of CNC milling machine bed using Al-SIC/graphite, al alloy and Al-SIC composite material, *Mater. Today Proc.*, 51, 735–741, 2021, doi: 10.1016/j.matpr.2021.06.219.

[20] H. Yu, D. Yu, C. Wang, Y. Hu, & Y. Li, Robotics and Computer-Integrated Manufacturing Edge intelligence-driven digital twin of CNC system: Architecture and deployment, *Robot. Comput. Integr. Manuf.*, 79, (February 2022), 102418, 2023, doi: 10.1016/j.rcim.2022.102418.

[21] M. Radovanovic, Multi-objective optimization of abrasive water jet cutting using MogA, *Procedia Manuf.*, 47, (2019), 781–787, 2020, doi: 10.1016/j.promfg.2020.04.241.

[22] C. Siemers, B. Zahra, D. Ksiezyk, P. Rokicki, Z. Spotz, L. Fusova, J. Rösler, & K. Saksl, Chip formation and machinability of nickel-base superalloys, *Adv. Mater. Res.*, 278, 460–465, 2011, doi: 10.4028/www.scientific.net/AMR.278.460.

[23] M. Rahman, W. K. H. Seah, & T. T. Teo, The machinability of Inconel 718, *J. Mater. Process. Technol.*, 63, (1–3), 199–204, 1997, doi: 10.1016/S0924-0136(96)02624-6.

[24] F. J. Sun, S. G. Qu, G. Li, Y. X. Pan, & X. Q. Li, Comparison of the machinability of titanium alloy forging and powder metallurgy materials, *Int. J. Adv. Manuf. Technol.*, 85, (5–8), 1529–1538, 2016, doi: 10.1007/s00170-015-8045-7.

[25] A. K. Parida & K. Maity, Hot machining of Ti–6Al–4V: FE analysis and experimental validation, *Sadhana – Acad. Proc. Eng. Sci.*, 44, (6), 1–6, 2019, doi: 10.1007/s12046-019-1127-8.

[26] A. J. Mian, N. Driver, & P. T. Mativenga, A comparative study of material phase effects on micro-machinability of multiphase materials, *Int. J. Adv. Manuf. Technol.*, 50, (1–4), 163–174, 2010, doi: 10.1007/s00170-009-2506-9.

[27] M. Quarto, G. Bissacco, & G. D'Urso, Machinability and energy efficiency in micro-EDM milling of zirconium boride reinforced with silicon carbide fibers, *Materials (Basel).*, 12, (23), 2019, doi: 10.3390/ma12233920.

[28]  D. Kim, S. R. Swan, B. He, V. Khominich, E. Bell, S. W. Lee, & T. G. Kim, A study on the machinability of advanced arc PVD AlCrN-coated tungsten carbide tools in drilling of CFRP/titanium alloy stacks, *Carbon Lett.*, 31, (3), 497–507, 2021, doi: 10.1007/s42823-020-00180-8.

[29]  B. Zhang, X. L. Zheng, H. Tokura, & M. Yoshikawa, Grinding induced damage in ceramics, *J. Mater. Process. Technol.*, 132, (1), 353–364, 2003.

[30]  H. P. Kirchner, Damage penetration at elongated machining grooves in hotcpressed Si3N4, *J. Am. Ceram. Soc,* 67, (2), 127–132, 2010.

[31]  A. Thakur & S. Gangopadhyay, Dry machining of nickel-based super alloy as a sustainable alternative using TiN/TiAlN coated tool, *J. Clean. Prod*, 129, 256–268, 2016.

[32]  P. Spriet, CMC applications to gas turbines, in N. P. Bansal & J. Lamon (eds.), *Ceramic matrix composites: Materials, modeling, and technology*, Hoboken, NJ: John Wiley & Sons, Inc, pp. 591–608, 2014, https://onlinelibrary.wiley.com/doi/book/10.1002/9781118832998

[33]  S. Yuan, Z. Li, C. Zhang, & A. Guskov, Research into the transition of material removal mechanism for C/SiC in rotary ultrasonic face machining, *Int. J. Adv. Manuf. Technol,* 95, (5–8), 1751–1761, 2017.

[34]  M. Hashish, A. Kotchon, & M. Ramulu, Status of AWJ machining of CMCS and hard materials, *Proceedings of InterTech. Conference, Indianapolis, IN.,* 1–14, 2015.

[35]  I. Tuersley & T. Hoult, The processing of SiC – SiC ceramic matrix composites using a pulsed Nd-YAG laser Part II the effect of process variables, *J. Mater. Sci,* 3, 963–967, 1998.

[36]  W. Li, R. Zhang, Y. Liu, C. Wang, J. Wang, X. Yang, & L. Cheng, Effect of different parameters on machining of SiC/SiC composites via pico-second laser, *Appl. Surf. Sci,* 364, 378–387, 2016.

[37]  C. Wei, L. Zhao, D. Hu, & J. Ni, Electrical discharge machining of ceramic matrix composites with ceramic fiber reinforcements, *Int. J. Adv. Manuf. Technol,* 64, (1–4), 187–194, 2013.

[38]  H. P. Zhang, Z. Zhang, Z. Y. Zheng, & E. Liu, Tool Wear in High-Speed Turning Ultra-High Strength Steel Under Dry and CMQL Conditions, *Integr. Ferroelectr.,* 206, (1), 122–131, 2020, doi: 10.1080/10584587.2020.1728633.

[39]  G. Kappmeyer, C. Hubig, M. Hardy, M. Witty, & M. Busch, Modern machining of advanced aerospace alloys-Enabler for quality and performance, *Procedia CIRP*, 1, (1), 28–43, 2012, doi: 10.1016/j.procir.2012.04.005.

[40]  H. Montazerian, A. Rashidi, A. S. Milani, & M. Hoorfar, Integrated sensors in advanced composites: A critical review, *Crit. Rev. Solid State Mater. Sci.,* 45, (3), 187–238, 2020, doi: 10.1080/10408436.2019.1588705.

[41]  B. Lauwers, Surface integrity in hybrid manufacturing processes, *Proc. Eng.,* 19, 241–251, 2011.

[42]  B. Lauwers, F. Klocke, A. Klink, A. E. Tekkaya, R. Neugebauer, & D. Mcintosh, Hybrid processes in manufacturing, *CIRP Ann.,* 63, (2), 561–583, 2014.

[43]  Z. Zhu, V. Dhokia, A. Nassehi, & S. Newman, A review of hybrid manufacturing processes-state of the art and future perspective, *J. Comput. Intergr. Manuf.,* 26, 596–615, 2013.

[44]  M. N. Mathabathe & A. S. Bolokang. Challenges in machining of advanced materials: Advances in sustainable machining and manufacturing processes, *CRC Press,* 2022, DOI: 10.1201/9781003284574-2

# 6 Characterization Techniques
## Material Testing and Characterization

*A. S. Bolokang and M. N. Mathabathe*

Multi-material additive manufacturing (MMAM) is a promising multi-functionality technique whereby tailored physical properties are achieved. For example, the 316L-HX multi-material is widely used for its excellent oxidation resistance in the chemical processing industry such as in retort components, furnace baffles, and support catalyst grids [1]. The emergent realization of MMAM is to evaluate the printability and characterization of the bond integrity [1].

Certainly, it is important to note that statistical methods for materials characterization and qualification is crucial to deduce data variables from inspection/testing of random samples. For example, the development and qualification of advanced gas reactor fuel programs require statistical acceptance tests. The statistical results require unambiguous interpretation methods to administer necessary measures of quality for a variety of materials [2].

## 6.1 THE DESIGN OF EXPERIMENTS AND OPTIMIZATION IN MANUFACTURING

Failure analysis of materials provide a solution to prevent machine down-time and production delays. Ductile to brittle fractures associated with shaft manufacturing, for example, will be delineated. In depth research, case studies, and further explanation on materials' properties/applications of common shafts will be outlined in this chapter.

### 6.1.1 Root-cause/failure analysis and testing of materials: Shaft examination

A shaft is a metal bar – usually cylindrical in shape and solid, but sometimes hollow – that is used to support rotating components, and transmit power and motion by rotary or axial movement. Bolts or studs can be considered stationary shafts, usually with acceptable tensile forces, but often combined with bending and/or torsional forces.

DOI: 10.1201/9781003356714-6 **113**

Shafts operate under a broad range of service conditions, involving corrosive and dusty environments that vary from extremely low, as in arctic or cryogenic environments, to extremely high, as in gas turbines. In addition, shafts may be subjected to a variety of loads such as tension, torsion, compression, and bending, or combinations of these. Shafts are also sometimes subjected to vibratory stresses [3].

Generally, shafts are common machine elements which are rotating members comprising a circular cross-section used to transmit power. A shaft may be hollow or solid, and for the purpose of power transmission it is supported on bearings and rotates a set of gears or pulleys. A shaft is generally acted upon by bending moment, torsion, and axial force. The design of shafts primarily involves determining stresses at critical points in the shaft that arises due to applied loading. Two other similar forms of shafts are axles and spindles. An axle is a non-rotating member used for supporting rotating wheels etc., and does not transmit any torque, while a spindle is described as a short shaft. However, design methods remain the same for axles and spindles as those for shafts [4]. Research studies have pointed out the importance of machine elements such as fasteners, springs, gears, valves, pipe fittings, etc., in engineering design. Jointly, shafts, are basic, important common machine elements. As an illustration, gears are generally mounted on some sort of shaft, and a gate valve or globe valve is opened and closed by a hand-wheel turning another type of shaft often referred to as a spindle [4].

Shaft materials are cost-effectively heat treated in order to provide good toughness combined with good strength. The commonly known specification is the UK En 19A, also known as the 709M40, and other alternatives include En 24, called 817M40, where 'clean steel' is a pre-requisite than the most commonly used material S132. Other grades are available where companies have developed specifications to overcome specific challenges associated with either mechanical properties or processing. OvaX 200 is one of a super-clean material to meet this need. This material has been designed to give the following listed below [5]:

- High cleanliness
- Slow air cooling
- Very low distortion
- Reduced grinding and finishing
- Temper resistance
- High operating temperature

The OvaX 200 is produced with a hardness of approximately 350 HB. The material is in an air-quenched & self-tempered condition containing a martensitic structure which can be machined directly with coated carbide cutting tools.

## Failures associated with shafts: Fracture origins
Fractures of shafts originate at points of stress concentration either inherent in design or introduced during fabrication or operation. Stress concentration design features include ends of keyways, press-fitted edge members, fillets, and edges of oil holes.

The major contributor to shaft failure is wear by bearings and fatigue. Fatigue occurs when an area of a material is a dynamically stressed area – typically a stress raiser, which may be mechanical, metallurgical, or sometimes a combination of the two. Shafts often break where high degree of stress concentration exists. Such stress concentration decreases fatigue resistance, especially when coupled with fretting. Metallurgical stress raisers may be quench cracks, corrosion pits, gross non-metallic inclusions, brittle second-phase particles, weld defects, or arc strikes etc. [3].

Appearance of most of common shafts, **Fig. 6.1**, include: (**a**) beach marks which indicate successive positions of the advancing crack front. The texture of the marks is usually smooth near the origin and becomes rougher as the crack grows; (**b**) ratchet marks which are numerous cracks that merge to form a single crack, and these marks are present between the crack origin; (**c**) chevron marks point to the origin of the crack. Rotational bending fatigue failures occur when each part of the shaft is subject to alternating compression and tension under a given load. A crack can start at any point on the surface where there is a stress raiser; (**d**) helical marks are torsional fractures with a 'twisted' display, which depend on the amount of torsional loading and whether the material is brittle or ductile. Assuming the shaft is ductile, it will show excessive twisting before failure; however, if it is brittle or subjected to extreme

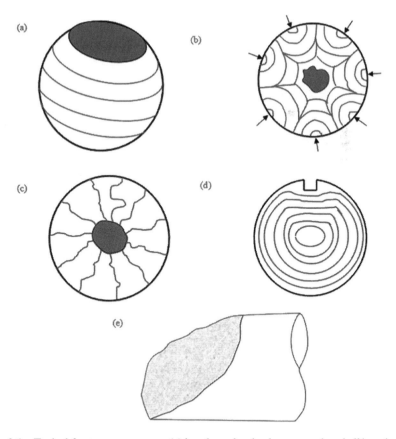

**FIG. 6.1** Typical fracture appearance: (a) beach marks also known as clamshell/conchoidal; (b) ratchet marks (radial steps); (c) chevron marks; (d) torsional (helical); and (e) brittle fracture

torsion, the fracture appearance will be rough; and (e) brittle fracture marks occur as a result of sudden torsional loading exhibiting a diagonal break with a rough surface.

## Examination of failed shafts

Most carbon steels utilized for shaft components are relatively ductile materials [4, 5]. However, shafts are manufactured with steps of different diameters along the length. These steps are stress raisers where a failure is more likely to occur. Even though the material is ductile, in the presence of a sharp radius at the step, the steel will act more like a brittle material and may fail before it reaches the maximum tensile strength [6]. Nyberg [6] explained that a break or fracture in a shaft is almost always initiated at some imperfection on the surface (stress raiser) at the tip of the crack. Depending on the type of loading used (tensile, torsional, and bending), the surface of the fracture reveals a clue to the magnitude of the load. This implies that if the appearance is brittle, then the failure occurred rapidly. On the other hand, if the failure is ductile or smooth, then the crack propagated for a long period before fracture.

## Design parameters

The performance of a gear set relies on the shafting for the gear elements, which should remain rigid to halt excessive deflection that would result in abnormal load distribution on the gear teeth. For the purpose of preventing failure, the fits between the shafts and the bearings and between the shaft and the mounted gears must be correct/precise. Additionally, strong shafts are also chosen because they can withstand fatigue loads that are added to transmitted torsional stresses and shock loads to prevent permanent yield [7].

## Mechanical conditions

Traditionally, fractures due to bending stresses are perpendicular to the shaft axis, whereas fractures resulting from fatigue-type torsional stresses are inclined at a 45° angle to the shaft axis. The fatigue cracks generally start at some stress raiser on the shaft such as sharp cornered fillets, snap ring grooves, fit corners, fretted areas, key or keyway ends, or tool or stamp marks [7]. *Stress raisers*: In cases where the fracture is caused by a stress riser that is symmetrical to the axis of the shaft, that is subjected to a low nominal stress, the crack usually starts from a single source and the final fracture area progresses from the surface towards the centre of the shaft. When the shaft is subject to high nominal stresses, many crack origins are found at the surface and the final fracture area is near the shaft centre. The fracture surface reveals the direction of rotation by locating the final fracture area with respect to the crack origin. *Keys and keyways*: Some fatigue fractures start from a combination of stress raisers such as a keyway and the end of a key. The strength of a keyed joint is affected by the distribution of the load along the keyway. The strength of the shaft at the keyway depends on position and accuracy of the key, the geometric shape of the keyway-end, and the relative location to the shaft shoulder fillets and the magnitude of the interference fit. A keyway fracture is normal to the shaft axis and/or it can peel circumferentially, which is common in a loosely fitted joint where nearly all the torque is transmitted through the key. Hitherto, fatigue starts at the bottom of the keyway and spreads around the shaft axis [7]. *Torsional failures*: Gear drive shafting failures are seldom due to pure torsion. However, when they do occur torsional cracks follow transverse or longitudinal shear planes, or diagonal planes of maximum tensile stress. For this reason, torsional fractures are more complex to analyze than bending fractures. Fracture due to a single overload in a ductile material may develop along the longitudinal shear plane, but in a brittle material the crack may develop on a 45° spiral angle perpendicular to the tensile stress. Torsional fatigue cracks may grow due to shear or tensile stresses or both. *Fretting corrosion*: Occurs when a little movement of tightly fitted parts occurs such as at bearing, gear, and coupling hub seats. As fretting progresses, small cracks appear at the surface of the shaft resulting in stress raisers which can become origins for fatigue cracks. The successful operation of a shaft can be halted by unexpected stress raisers formed in the processing of a shaft or during the operation of the gear drive. There may be other situations where an unexpected dynamic load is imposed on the shaft which can be two to three times greater than the operating load [7]. *Abrasive wear*: Also known as abrasion, this is caused by the displacement of material from a solid surface because of sliding hard particles adjacent to the surface. Abrasive wear affects the size and shape of a shaft. Some examples of abrasive wear of shafts are foreign particles such as sand, dirt, metallic particles, and other debris in the lubricant [8]. *Fatigue failures*: Commonly begins at stress raisers such as arc strikes, localized corrosion, shrink fits, grooves, quench cracks, welding defects, notches etc. Generally, failures associated with fatigue fractures may occur due to misalignment caused by the mismatch of mating parts. Misalignment can be introduced during original assembly of equipment and after a repair of a piece of equipment. Deflection or deformation of supporting components in service may also cause misalignment which causes vibrations resulting in a fatigue failure of the shaft [8].

## 6.2  THE NECESSITY AND CLASSIFICATION OF MATERIAL TESTING

Mechanical testing is categorized into static, dynamic, and fatigue testing, as a function of time. In static tests the load on the test-piece is increased gradually or maintained constant for a long time resulting in strain rates that are very low. In the case of dynamic tests, where impact loadings are prevalent, the test-piece is loaded at substantial speeds so that the strain rate is high. Meanwhile, in fatigue tests, the test-piece is put through repeated loading which varies either only in magnitude or in magnitude and direction [9]. Practices of load application in mechanical tests differ i.e., tests in tension, compression, torsion, bending etc., while mechanical strength of materials can be ascertained either at room temperature or intermediate to high temperatures (viz.: impact-bend, creep, rupture strength tests etc.), controlled by the service status quo of the tested metal.

Atomic simulations in materials science are computational systems which permit the performance of large-scale modelling at a fairly low cost [10]. Bonding between atoms is defined by the use of empirical or semi-empirical interdependent potentials [10]. Alternatively, ductility advancement of $\gamma$-TiAl-based alloys is via ternary, quaternary and quinary microalloying elements [11–13]. For example, TiAl intrinsic ductility is formed due to directional Alp-Tid covalent bonds. Further, the addition of ternary/quaternary elements reduces the p-d reciprocal action and completely improves the ductility of TiAl alloys. Considering that TiAl lattice tetragonality is related to inadequate ductility [14], we previously investigated the combined effects of these quaternary additions on the c/a ratio of γ-TiAl-based alloys. Thus, we reported on the influence of Mn, Nb, Sn, and Si alloying on the γTiAl from the preferential site occupancy, interfacial energetics to its physical properties, and their corresponding experimental validation, using density functional theory (DFT) equipped with Cambridge Serial Total Energy Package (CASTEP) [15]. The physical properties determined were the Poisson's ratio, hardness, universal anisotropy, bulk modulus, shear modulus, and Young's modulus. The results indicated that all the hardness values were greater than 5HV. The universal anisotropy ($A^U$) [16] of the structures were calculated and ranged from 1.34, 2.99, to 4.4, for the L10-TiAl, Ti11Al5, and Ti8.3Mn0.7Nb2Al3.7Si0.3Sn, respectively. The properties obtained prompted the elaboration of: (1) internal characteristics constituting resistance anisotropy of slip systems, (2) the lamellar colony size of 100 μm–1 mm and lamellar thickness of 1–10 μm, and (3) Burgers orientation relationship (BOR) [17]. It's worth mentioning that the symmetrical crystal universal elastic anisotropy ($A^U$) as a criterion to ascertain mechanical properties was exploited in the study:

$$A^U = 5\frac{G_V}{G_R} + \frac{B_V}{B_R} - 6 \tag{6.1}$$

where GV and GR are the Voigt and Reuss shear modulus, and BV and BR are the Voigt and Reuss bulk modulus, respectively. The crystallization of (L10-TiAl) tetragonal structure stem-forth anisotropy factor is equal to unity (AU~1) [18]. The Voigt-Reuss-Hill (VRH) averaging scheme [19–21] was employed to calculate mechanical properties such as (1) Young's modulus (E) – a measure of elasticity; (2) Shear (G) – capacity to resist deformation under shear stress; and (3) Bulk (B) modulus – resistance to deformation under applied stress [22]. In addition, the results showed that the ductility condition was complacent with L10 (G/B = 0.55), denoting that the addition of Mn, Nb, Sn, and Si to the Ti11Al5 contributed to ductility with a quasi-lower shear modulus, and Poisson's ratio of <0.35 [15].

Non-destructive testing and evaluation (NDT & E), is the science and technology of assessing materials, and examining of discontinuities in components, and/or dissimilarity in features without damaging the serviceability of the part or system. **Figs. 6.2** and **6.3** highlight some of the instrumentation used for NDT inspection of materials.

**Fig. 6.2** shows an industrial radiography machine which is an element of NDT. It is a method of inspecting materials for hidden flaws that would otherwise be invisible.

Phased-array ultrasonic testing **Fig. 6.3** can be used in a variety of applications and industries. As a non-destructive testing method, it is more reliable, more effective, and faster than many other traditional methods, such as radiographic inspection. In addition, some benefits include improved portability, convenience, inspection speed, and safety.

**Fig. 6.4** one of many electromagnetic testing techniques which employs electromagnetic induction to detect and characterize surface and subsurface flaws in conductive materials.

**FIG. 6.2**   Gamma inspection and X-ray equipment

**FIG. 6.3**   NDT equipment for phased-array ultrasound

**FIG. 6.4**   NDT equipment – Eddy current testing (ECT)

**Fig. 6.5** Is an NDT MPI testing method for detecting surface and slightly sub-surface discontinuities for ferromagnetic materials, while Dye Penetrant Inspection (DPI), is a widely applied and low-cost inspection method used to locate surface-breaking defects in all non-porous materials such as metals, ceramics or plastics.

Infrared thermography **Fig. 6.6** is an NDT method involving the induction of heat flow in a test component by internal conditions and measured on the surface by an InfraRed camera. This technique detects not only the smallest surface defects, but also internal structural defects under the surface.

**Fig. 6.7** is NDT Ultrasonic Testing (UT) equipment based on the propagation of ultrasonic waves in the material tested. In the most common UT applications, very short ultrasonic pulse-waves with centre frequencies ranging from 0.1 to 15 MHz, and occasionally up to 50 MHz, are transmitted into materials to detect internal flaws or to characterize materials.

An Ultrasonic Thickness Gauge (UTG), **Fig. 6.8**, is an NDT method that evaluates the thickness of a component by measuring the time it takes for sound to travel from

**FIG. 6.5** NDT equipment, magnetic particle inspection (MPI)

**FIG. 6.6** Some NDT equipment Infrared thermography

FIG. 6.7   Some NDT equipment Ultrasound testing (UT)

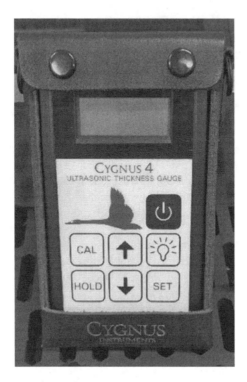

FIG. 6.8   Some NDT equipment Thickness gauge

the transducer through the material to the back end of a component, and then back to the transducer. The gauge ascertains the thickness based on the sound velocity through the material being tested. The standard frequency used by an ultrasonic thickness gauge is 5 MHz. UTG are generally used for corrosion monitoring, wall thickness measurement, and casting/moulding thickness testing.

## 6.3   IMPORTANT INSTRUMENTS FOR TESTING AND CHARACTERIZATION

The X-ray powder diffraction (XRD) is a rapid analytical technique primarily used for phase identification of a crystalline material and can provide information on unit cell dimensions. The analyzed material is finely ground to a flat surface, and the average bulk composition is determined. X-ray diffraction is based on constructive interference of monochromatic X-rays and a crystalline sample. These X-rays are generated by a cathode ray tube, filtered to produce monochromatic radiation, collimated to concentrate, and directed toward the sample. The interaction of the incident rays with the sample produces constructive interference (and a diffracted ray) when conditions satisfy Bragg's Law ($n\lambda = 2d \sin \theta$) [23]. It is worth noting that there is a significant difference: $|\Delta\ dS^{elect}\ (s,\ t) - \Delta dS^{xray}\ (q,\ t)$ between the X-ray and the electron diffraction signals [25], whereby $s$-is the momentum transfer in electron diffraction; then $q$-which is the momentum transfer vector (difference amid the incident and scattered wave vectors) in x-ray diffraction; and $t$ is the interval time between pump and probe pulses [24]. Certainly, the non-negligible difference between the two diffraction signals is due to the contrasting dissimilarity which is attributable to the supplementary impact from the nuclei. The difference is miniature in the small-angle scattering, but gets larger at intermediate angles up to 11 Å$^{-1}$, correlative to elevated resolution [24].

Neutron scattering (NS) is a characterization technology designed to resolve microstructural properties across a range of scales, with its high penetrating capacity attributable to analysis of confined-fluid behaviour. Neutrons are stewards for exploring the nuclear and magnetic structure of condensed matter and lattice dynamics. Further, since neutrons are electrically neutral and pierce into a sample, they can be utilized to analyze large rock samples [25]. Small-angle neutron scattering (SANS) samples are conducted using three configurations where distance detector-sample, wavelength, and medium angle and wide angle configurations apply [26].

Mössbauer effect spectroscopy (MES) is appropriate for the study and detection of carbides. In most cases the method ascertains the transition of carbides for quenched and tempered materials, particularly Fe-C steels with C content in the range of (1.2–1.6wt.%) [27]. Position annihilation spectroscopy (PAS) is an excellent tool for examining materials and their defects. However, the method poses a challenge in the analysis of bulk materials due to the limited range of positrons in matter; thus restricted to a thin sample analysis [28].

The most widely used method in catalysis is electron paramagnetic resonance (EPR). It is used mainly in the characterization of electron structures and to demonstrate the nature of active sites in anion-doped based materials. The applications of titanium and its alloys have recently been incorporated into this approach [29].

Surface enhanced Raman scattering (SERS) is a surface characterization tool, which is also a vibrational spectroscopy technology. This method elevates the electromagnetic field produced by local surface plasmon to conduct sensitive structures on low-concentration analytes [30]. Raman scattering is a dominant and non-destructive technique used to evaluate the vibration modes of the microstructure [31]. Raman spectra collected along the reaction product surface layer after cold-pressing of Ti-48Al (binary), Ti-48Al-2Nb (ternary), and Ti-48Al-2Nb-0.7Cr (quaternary) metallic powder alloys revealed varied features illustrated in **Figs. 6.9** and **6.10**, which compares the Raman spectra of the cold-pressed binary, ternary, and quaternary powder mixture with those of heated alloys at 1400 °C. The cold-pressed binary powder mixture TiAl-based alloys in **Fig 6.9** exhibited three Raman active modes at 841 cm$^{-1}$, 1006 cm$^{-1}$, and 1036 cm$^{-1}$. The Raman peaks at 1006 cm$^{-1}$ and 1036 cm$^{-1}$ appeared as one peak split into two, which is a representation of a double in this range. Like the binary (TiAl) compressed powder mixture, the same Raman peaks at 1400 °C were observed in the ternary and quaternary cold-pressed alloy powders, with an additional strong Raman peak at 1372 cm$^{-1}$. These peaks could be attributed to the evolution of the rhombohedral orientation detected in alloys containing BCC elements, as observed in our previous study [32]. The Raman peak at 828 cm$^{-1}$ and 1033 cm$^{-1}$ are Ti related and slightly smaller to those obtained in pure Ti after water quenching [33, 35]. The cast/heated binary, ternary and quaternary alloys in **Fig. 6.10** had only two peaks at 826 cm$^{-1}$ and 1038 cm$^{-1}$. The broad peak at 828 cm$^{-1}$ was closer to that observed at 600–750 cm$^{-1}$ on Ti$_2$AlC due to Ti-C bonding vibration [36, 37].

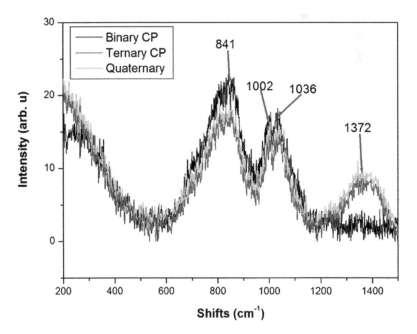

**FIG. 6.9** Raman spectra of binary, ternary, and quaternary cold-pressed powders performed at room temperature

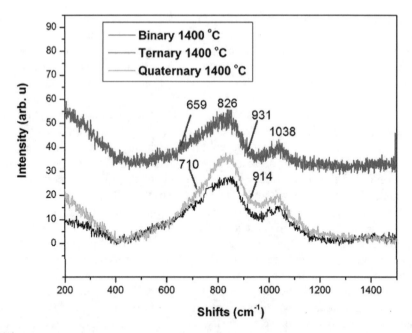

**FIG. 6.10**   Raman spectra of binary, ternary, and quaternary cold-pressed powders performed at 1400 °C

On the contrary, the large broad hump at 826 $cm^{-1}$ may be related to Al-Nb bonding, where Ti seems to have been etched away by the compression mechanism on the surface layers confirmed by EDX analysis performed in [32]. Therefore, this Raman feature may be connected to the imperfection of the γ-TiAl-X $_{(Nb, Cr)}$ lattice in the studied alloys.

Infrared spectroscopy (IR) can be coupled with other techniques as far as characterization is concerned. In the work by Luo et al. [38] it was equipped with atomic force microscopy to ascertain IR absorption spectra and images with linear resolution of 50–100nm, of microplastics (MPs). Particle characterization by light scattering has been used in many fields such as meteorology, cosmology, oceanography, and nanotechnology. Light scattering occurs with a variety of particles sizes, shapes and origins in different environments. There are numerous light scattering technologies for particle characterization, e.g. the laser diffraction (LD) method [25].

Atomic force microscopy is an important tool for the characterization of the nanomechanical properties of any material. The measurement of atomic force is conducted using a dimension instrument in contact mode, employing the Au reflective silicon-nitride coating tips with a constant spring of $0.07Nm^{-1}$ and probe radius of 20nm [39]. Mass spectrometry is one of the metallomic approaches used in the development of metal-based anticancer drugs. Generally, this has been achieved by using inductively coupled plasma-mass spectrometry (ICP-MS) for investigations of metal-based heterogeneous and structural and functionally complex solid tissues [40]. Ion beam analysis

(IBA) is a well-known approach for material characterization where interest lies on near or surface regions. This analytical technique comprises of advanced algorithms and synergistics for multiple measurements. Statistical methods fitted in the work by Silva et al. [41] proved to be resilient and crucial for quantitative classification.

Over the past 20 years, neutron activation analysis (NAA) has been employed for obtaining various nuclides in environments containing phosphate rocks and soils of different chemical composition. Research has shown many attempts to assess the effectiveness of NAA by achieving lower limits of detection. It was shown that interfering elements that exhibit higher backgrounds may possibly enhance the detection limits [42].

For any interface chemistry determination, such as pore size distribution, specific area, metal speciation, and other features, SEM/EDAX can be utilized. For example, characterization techniques viz., gas adsorption porosimetry and SEM/EDAX were employed concurrently to analyze the properties of desorption reactions of manganese ions on the coated sand surface in water [43].

On the other hand, electron backscatter diffraction (EBSD) analysis is a very powerful tool for microstructural characterization. It is a scanning electron microscope (SEM)-based technique that gives crystallographic information about the microstructure of a sample. The common way to distinguish EBSD is its colour representation which is used as a tool to visualize local crystal orientations, which are expressed in Euler angles ($\varphi_1\phi\varphi_2$), which is, therefore, a representation of successive conventional rotations to match the crystal orientation with a Cartesian coordinate system related with the sample surface [44]. For micro/macro texture determination, electron backscatter patterns are generated by the interaction of the primary electron beam with a tilted specimen, known as the electron backscatter diffraction patterns (EBSP). To view the diffraction patterns, there are two conventional prerequisites of topographical information necessary for SEM operation. The first requirement is that the angle between the incident beam and the specimen surface must be ~10–30 ° to reduce the signal absorbed, and to maximize the diffraction proportion. The second requirement is related to the specimen preparation, i.e. beam/specimen interaction depth of ~10nm, which is strain free and clean, and where electrical conduction is amended by deposition of a thin layer of carbon allowing an ease of patterns attainment [45]. The EBSP's are essentially Kikuchi patterns, and they occur in the SEM when the stationary probe is fixated on the specimen. The subsequent elastic scattering of the divergent electrons by the crystal planes configures an apparel of Kikuchi cones when the Bragg condition is satisfied [45]. If the distribution of diffracted intensities is recorded by the interception of Kikuchi cones by photographic film or a recording screen, the resulting EBSP can be viewed as shown in **Figs. 6.11** and **6.12**, which is the L10-TiAl or γbulk phase of Ti-48Al-2Nb-0.7Cr-0.3Si alloy. **Figs. 6.13** and **6.14** are examples of EBSD analysis of a cold-rolled AA6016 Al alloy sample up to 2mm thickness. **Fig. 6.13** demonstrate an inverse pole figure (IPF) map with its corresponding pole figure maps for the {111} planes which demonstrated profound texture indices at maximum intensity of ~ 3 and 8 for the Al solid solution and Mg2Si-β/eutectic-Si, respectively. Another example of generating textures is in the form of macrotexture orientation distribution function (ODF) shown in **Fig. 6.14**

**FIG. 6.11**   EBSD analysis showing Kikuchi pattern/bands of γ-TiAl phase

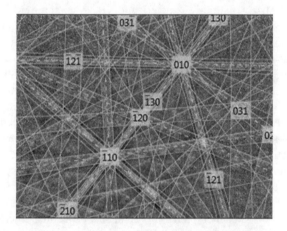

**FIG. 6.12**   EBSD analysis showing pattern indexation of γ-TiAl phase

indicating strong fibre textures at densities of $\phi 2 = 0°–90°$ serial sections with black contour lines.

It has been found that the transmission electron backscatter diffraction (T-EBSD), coupled with the SEM-based electron diffraction characterization technique, may possibly contribute enhancement in spatial resolution over common EBSD with detailed quantitative orientation data matching that attained by transmission electron microscopy (TEM). This was investigated by Sun [46], where more focus was on the recrystallization analysis of the dislocation sub-microstructures of the γ′ precipitates. It was discovered that dislocation motion transfers from a pile-up at the matrix /γ′ interface into the primary γ′precipitates [46].

**Table 6.1** shows some of the commonly used instruments for characterization of materials.

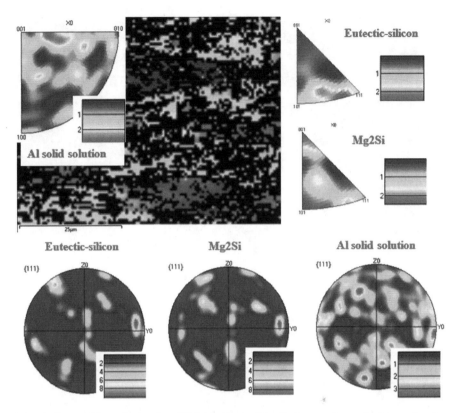

**FIG. 6.13** EBSD analysis of a 6016 Al alloy cold rolled up to 2mm thickness showing inverse pole figure map with their corresponding pole figures

### 6.3.1 SOME THERMAL CHARACTERIZATION TECHNIQUES

Thermal analysis measurement methodology, also known as double patterning (DP), was developed in Reference [64] to validate the possibility of achieving an optimal curing temperature for sub-40nm half-pitch devices. The approach exhibited the use of differential scanning calorimetry (DSC), thermogravimetry analysis (TGA), and dynamic mechanical analysis (DMA) techniques, which therefore provided the determination of an accurate curing temperature which contributed to maximum crosslinking and least amount of resistance degradation [64]. **Fig. 6.15** is an image of DSC-TGA equipment.

Magnetic materials play a fundamental part in the energy conversion in electrical machines. Generally, experimental setups are proposed to characterize the electromagnetic properties as a function of temperature up to ~200 °C. In reference to this, Jamil et al. [65] studied temperature dependence of electromagnetic properties such as iron losses, electrical conductivity etc. during the operation of Claw Pole (CP) machines. The findings indicated that at elevated temperatures, the saturation magnetization, material permeability, and coercive field decrease. Eminently,

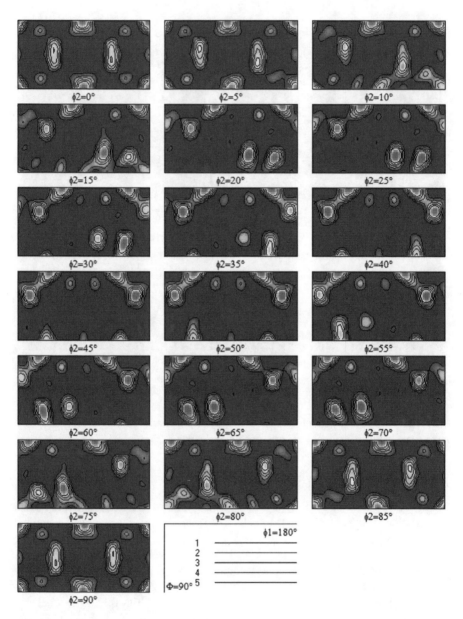

**FIG. 6.14** EBSD analysis of a 6016 Al alloy cold rolled up to 2mm thickness-Serial section through an orientation distribution as per Fig. 6.13

**TABLE 6.1**
**Some examples of advanced techniques for materials characterization**

| Type of characterization technique | Material and/or information characterized | Application | Conclusions | References |
|---|---|---|---|---|
| **Diffraction techniques** | | | | |
| X-ray diffraction (XRD) | • Multiphase Bi-Sr-Ca-Cu-O (BSCCO) thin films | High temperature where superconductivity is required | XRD analysis obtained volume fraction of the phases i.e., 24 & 30 Å c-axis oriented increased to the 37 Å relative to BSCCO-MgO interface. | [47] |
| Neutron scattering | • Graphite vs LiNi0.33Mn0.33Co0.33O2 pouch cell | Battery systems | Correlation of compositional information to a specific size range was established. | [48] |
| Small-angle neutron scattering (SANS) and small-angle X-ray scattering (SAXS) | • Two lipid system viz.: Soybean oil and Mygliol 812 | Pharmaceutical applications for cancer side effect reduction | Both techniques revealed structural information e.g. cubic structure with polydispersed non-fractal objects that varies with concentration | [49] |
| **Spectroscopic techniques** | | | | |
| (IR) | • Mixture of $H_3BO_3$, $Fe_2O_3$, $NaCO_3$, and $NH_4H_2PO_4$ powders | Electrical and optics applications | The overall network structure was determined. The Poisson's ration constancy shows that there is no change in cross-link density. | [50] |
| Raman spectroscopy | • Ba0.97La0.02TiO3 polycrystalline ceramic | Enhanced electrical properties are applicable | The technique showed that the dopants of $La_3+$ and $Nb_5+$ in the A- and B- sites, confirms a disorder in the structure. | [51] |
| Electron paramagnetic resonance (EPR) | • Duplex stainless steel | Oil and gas industries | The use of solubilized condition of steel as reference value for one site testing is preferred | [52] |

*(Continued)*

**TABLE 6.1 (Continued)**
**Some examples of advanced techniques for materials characterization**

| Type of characterization technique | Material and/or information characterized | Application | Conclusions | References |
|---|---|---|---|---|
| Positron annihilation spectroscopy | • $Bi_2O_3$ modified titanium sodium lead borosilicate glass | Optics and radiation shielding applications | The physical properties of glasses were improved by incorporating $TiO_2$ | [53] |
| Mössbauer spectroscopy | • $MFeOPO_4$ (M:Co, Ni) materials | Battery systems | New insight about the reaction mechanism of the material was acceptable. Additionally, $Fe^0$ was discovered at the end of discharge at 0.01V | [54] |
| **Compositional characterization techniques** | | | | |
| Mass spectrometry | • $TiO_2$ nanoparticles | Biological medicine | All Ti ions and their oxide ions were spotted and imaged. | [55] |
| Neutron activation analysis | • Iron and chromium | Nuclear and aerospace applications | Measurements conducted on the given materials, and the obtained mass detection limit (MDL) were 71g and 88g, respectively. | [56] |
| Microanalysis by electron beam | • High-purity tungsten single crystal | High temperature, strength, elastic modulus, and density applications | The results indicated the dependencies of flow stress and its strain-rate sensitivity on temperature. | [57] |
| Ion beam analysis | • Extruded Al-Mg-Si-Cu alloys | Light weight applications | Characterization of an α-Al(Fe, Mn) Si dispersoid indicate the segregation at the phase boundary and in the shell of the dispersoid. | [58] |

**Synchroton and surface techniques**

| Technique | Material | Application | Findings | Ref. |
|---|---|---|---|---|
| Synchrotron radiation | • Zn-27Al-3Si-xSr alloys (x=0%, 0.06%, 0.1%, 0.2%) | High temperature, wear resistance alloys, and good dimensional stability are applicable | Sr had an effect on spheroidizing the primary Zn-27Al-3Si alloy, by generating the high density of twins resulting in alteration of the growth process of primary Si. | [59] |
| X-ray photoelectron spectroscopy | • Multiple oxide comprising NiO, $TiO_2$, & $NiTiO_3$ | Solar-heat shielding materials, coating materials for amenities, semiconductors, and photocatalyst | Calculated $[NiTiO_3]/[NiO]$ were corresponding with theoretical ratios. | [60] |

**Microscopic techniques**

| Technique | Material | Application | Findings | Ref. |
|---|---|---|---|---|
| Atomic force microscope (AFM) | • 316L steel powders | Where recyclability of leftover metallic powders is applicable | ~10% more pores in recycled powders which indicated an increase in surface roughness, with subsequent reduction in hardness and modulus | [61] |
| Particle characterization by light scattering | • Polystyrene latex particles | Dielectric properties applicable | Optical chambers provide precise measurements allowing good conductivity and frequency range. | [62] |
| Transmission Electron Microscopy (TEM) | • Cu-10wt.%Co magnetic materials | Higher storage density of magnetic devices (Magnetic nanofields) | It was measured that phase shift in the in-plane component of the magnetic field in cobalt is 1.2T on average. | [63] |
| Scanning electron microscope-electron backscatter diffraction (SEM-EBSD and energy dispersive X-ray (EDAX) | • AISI 304 & 316 STEELS | Corrosive environments | The welded materials performed better with improved mechanical properties | [64] |

**FIG. 6.15**   DSC/TGA instrument

thermochemical heat-storage is a favourable technology for use in renewable energy reserves. Other examples, including mixed-salt porous-carrier composites, have been fabricated and characterized for application in energy storage systems. However, there are drawbacks experienced during the process, and in the work by Korhammer et al. [66] the disadvantages of calcium chloride ($CaCl_2$) in thermal energy storage was overcome by the addition of potassium chloride (KCl), enabling water uptake of two time higher than that of $CaCl_2$ during the hydration process. Silica gel-supported lithium chloride composite sorbents regarding water uptake embedded in silica gel pores was examined in [67]. The findings indicated that impregnating silica gel in the 30wt.%LiCl solution resulted in optimum stable storage performance [67].

A novel method for thermal characterization of homogenous materials has been developed by Barrio et al. [68, 69] using the Karhunen-Loève decomposition (KLD) technique. The method showed two advantages, viz. (1) KLD can be employed even when analytical solutions of heat conduction is not provided, and (2) dispensation of minimized noise allowing good prediction even when poor quality data is provided, while the same method was employed for heterogenous materials [70]. Efficiency of energy systems that involves cold energy can be simulated using cold thermal energy storage (CTES) [71] whereby the findings are in good agreement with experimental results, and the fundamental boundary conditions satisfied.

### 6.3.2   Some metallic powder characterization techniques

Knowledge of metallic powder comportment is complex, particularly when subjected to processing at various physical environments. Powder flowability, or when the metallic powder flows from the hopper into the dies, typically determines hardness, weight, and content uniformity, and is essential for blending. There are several methods of measuring powder flow, involving bulk density, tap density, compressibility index, angle of repose, or Hausner's ration [72]. A powder rheometer called FT4 (**Fig. 6.16**) is another method designed to characterize powders at different conditions. Powder

**FIG. 6.16** FT4 powder rheometer

particles with low internal porosity and a high density exhibit free-flowing properties. However, inherent surface roughness is attributed poor flow characteristics as
a result of friction and cohesiveness, and the following points apply: (1) effect of
particle size – fine particles with increased surface to mass ratio are more cohesive
up to 250μm particle compared to coarser particles [73]; (2) effect of particle shape –
similar powder sizes but with dissimilar shapes have significant difference in flow
properties [74]; (3) density of particles – more dense particles are less cohesive when
compared to less dense particles; (4) potential for electro-static charge – handling
and processing may lead to electro-statically charged particles resulting in changes in
powder behaviour; and (5) adhesion and cohesion – solid particles stick to themselves
or other surfaces due to the presence of molecular forces. Cohesion generally exists
due to like surfaces, while adhesion exists between two unlike surfaces.

The benefits of applying cold spray to compacted metallic powders is the absence
of chemical reactions or phase alterations during the deposition process. The other
important benefits include low oxygen content within the material, relatively high
deposition rates (10–30kg/h) and avoided shrinkage on cooling when compared to
thermal spraying [75]. Cold spraying provides inter-particle bonding that is strong
such that the formation of pores will be prevented when the component shrinks.
However, if the bonding is weak, the interface will open-up generating pores known
as the Kirkendall effect, which can be counteracted by plastic flow and creep [75]. As
the initial intermetallic phase emerges, shrinkage takes place. As a result, interfaces
between particles become widely open, creating pathways for residual oxygen to
travel from the samples' surface to their centres, causing oxidation along metal grain
boundaries and the loss of the grains [76]. Since oxygen is a contributing factor for
the mentioned issues, generally, Leco analyzers are used for the determination of
interstitial elements such as carbon, hydrogen, oxygen etc. **Figs. 6.17** and **6.18** show
examples of Eltra analyzers.

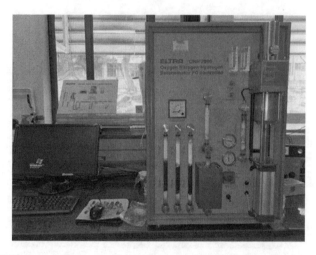

**FIG. 6.17**   ELTRA analyzer equipped with PC controllers for oxygen, nitrogen, and hydrogen determination

**FIG. 6.18**   ELTRA analyzers equipped with PC controllers for carbon, and sulphur determination

Thermal characterization techniques (**Table 6.2**) are generally conducted on advanced materials to gauge how they perform in service temperatures, which is of course dependent on their applications.

## 6.4   MICROSTRUCTURE CHARACTERIZATION

Materials characterization is a pivotal process in which a material's chemical, microstructure, and physical properties are canvassed, quantified, and examined using a variety of analytical approaches, techniques, and tools. **Sections 6.4** and **6.5** mainly focus on the case studies implementing the methods delineated in **Section 6.3**.

**TABLE 6.2**
**Thermal characterization techniques**

| Fabrication method | Materials | Application | Thermal characterization technique used | Comments | Reference |
|---|---|---|---|---|---|
| Laser cladding | MMC 316L stainless steel + WC | Metal coatings | Differential thermal analysis (DTA) | DTA employed to achieve complete remelting of the clad composite | [77] |
| DCN and TEOS | Siloxane copolymer with silicate hybrids | Electronic, solar, and optoelectronic material industry | DSC | Thermal transitions with no decomposition below 200 °C were observed | [78] |
| Titanium dioxide production industry | Inorganic TiO2 waste materials i.e. red gypsum and ilmenite | Fire wall insulation, or fire-resistant panels | TGA | Thermal behaviour showed little chemical and structural changes up to 1000 °C | [79] |
| Simulative sorption experiment | Vermiculite/SrBr2 | Renewable energy systems | STA, TGA and DSC | Heat released in salt dissolve process reduced as the salt concentration decreases | [80] |
| Hot- box method | Transparent porous aerogel | Next-generation window insulation application | Reduced-scale hot box (RHS) validated by laser flash method | The thermal conductivity of porous window material was ~0.018W/ mK at room temperature with 50% reduction of thermal transmittance coefficient | [81] |

MMC = metal matrix composite; WC = tungsten carbide; DCN = 4-[[5-dichloromethylsilyl]pentyl]oxy]cyanobenzene; TEOS = tetraethoxysilane, DSC = differential scanning calorimetry; TGA = thermogravimetry analysis; STA = simultaneous thermal analyzer

### 6.4.1 Case Studies in Material Characterization

Two case studies are presented here to illustrate the characterization challenges that occur in (1) deformed 6016 Al alloy used in the automotive industry – innovative design process strategies came together for successful fabrication of the alloy; and (2) providing solutions to root-cause analysis of a broken shaft during service.

### Case study 1: Texture analysis of cold-rolled 6016 Al alloy in different thickness reduction i.e., 20 and 60%

Characterization of a rolled 6016 Al alloy via compositional analysis using XRD and texture analysis using SEM-EBSD are examined in this case study.

#### *X-ray diffraction*

The XRD patterns in **Fig. 6.19** denote rolled 6016 Al alloy in 20% and 60% respectively. The major peaks were successfully identified as al ss which comprises Al8Mg, eutectic-Si, and $Mg_2Si$ composition. The $Al_8Mg$ prototype $MgZn_2$ crystallizes in fcc cubic structure with an Fd3m 227 space-group atoms in the unit cell with lattice parameters a = 28.6Å. The XRD spectrum of rolled 20% thickness reduction alloy at 17.191 2θ °detected a cubic Fm3m 225 phase with a = 10.3080Å; this is represented by the insert in **Fig. 6.19a**, while after 60% thickness reduction during rolling, the phase seemed to have disappeared for the alloyed rolled at 60% reduction.

Moreover, the diffraction patterns of the alloys were analyzed by simulating line profiles employing the Williamson-Hall (W-H) method (**Figs. 6.20** and **6.21**). The

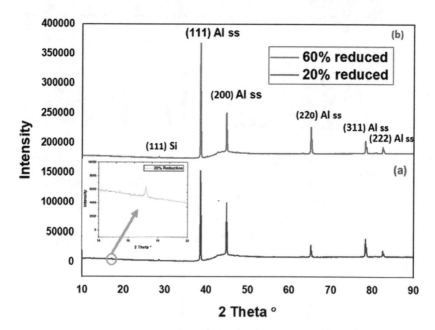

**FIG. 6.19**   XRD patterns of 6016 alloy rolled in (a) 20% and (b) 60% thickness reduction. *Al ss stands for the aluminium solid solution

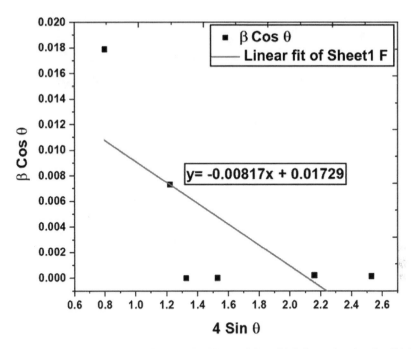

**FIG. 6.20**   The W-H plot examining the FWHM peak breadth information for the 6016 Al alloys rolled at 20 percentage thickness reduction

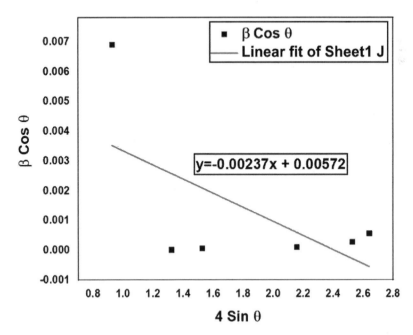

**FIG. 6.21**   The W-H plot examining the FWHM peak breadth information for the 6016 Al alloys rolled at 60 percentage thickness reduction

**TABLE 6.3**
**Crystallite size and microstrain from XRD data using W-H plot**

| Thickness reduction | Crystallite size D (nm) | Strain (ε) |
|---|---|---|
| 20% | 8 | 0.00817 |
| 60% | 58 | 0.00237 |

W-H plots were attained by pattern fitting, using full-width-half-maximum (FWHM) or β values obtained from the peaks. The average-size information was obtained using intercept of the outstanding fit linear regression in the W-H plot, meanwhile, the slope was utilized to attain the strain of the system. The results βCos θvs 4 Sin θvalues show that both the alloy planes rolled in different percentage reductions exhibited significant negative slopes on fitting which demonstrates that the microstrains of 60% reduced alloy during rolling cannot be a predominant source of peak broadening. On the contrary, the relative non-linearity in the 60% reduced alloy during rolling may suggest a rather slight non-homogeneity in particle/grain size, evidenced by an increase in crystallite size D = 58nm, compared to its counterpart (**Table 6.3**). The results show that with decreasing percentage reduction during rolling of 6016 Al alloy, the crystallite size decreases. For example, the 20% thickness reduced rolled alloy exhibiting a D = 8nm may be related to the accumulation of lattice strains in the nano-grain boundaries due to the detected phase circled in **Fig. 6.19**, with the resultant improvement of Gibbs free energy thereafter.

*Electron backscatter diffraction (EBSD)*

The EBSD analysis of both the rolled alloys in varying thickness reduction i.e., 20 & 60%, are shown in **Figs. 6.22** and **6.23** respectively, indicating longitudinal grains as demonstrated by the dashed lines (**Fig. 6.22d**) insert. **Fig. 6.22** is the grain orientation of rolled 20% reduced 6016 Al alloy, showcasing an electron image (**Fig. 6.22a**), Euler angle grain map (**Fig. 6.22b**), inverse pole figure maps (**Fig. 6.22c-e**), and phase colour map (**Fig. 6.22f**). The electron image insert in **Fig. 6.22a** is analyzed in a longitudinal view, indicating eutectic-silicon particles in an Al ss matrix, analogous to **Fig. 6.23g** composition. As it can be observed by the EDX spectrum, C and/or O are negligible, which may be detected because of carbon tape. Correspondingly, its IPF map is shown by the insert in **Fig. 6.22d**. It is important to note that the EBSD indexing also detected $Mg_2Ni_3P$, a rhombohedral phase with a space group of R-3M 166, and a = b = 4.9710 Å, c = 10.9610 Å for both the alloys.

**Fig. 6.23a** demonstrates an electron image of rolled 6016 Al alloy at 60% reduction, indicating eutectic-silicon particles whose composition is shown in **Fig. 6.23g**. The Euler map is shown in **Fig. 6.23b**, while the IPF maps are displayed in **Fig. 6.23c–e**, and **Fig. 6.23f** is the colour map. It is clear from these results that the Al ss comprised of eutectic-Si and $MgZn_2$ were profound for this indexing, while the volume fraction of the remaining phases were minimal. The phase acquisition for the alloys is justified in **Table 6.4**, and the IPF maps for both the alloys showed relatively recrystallized

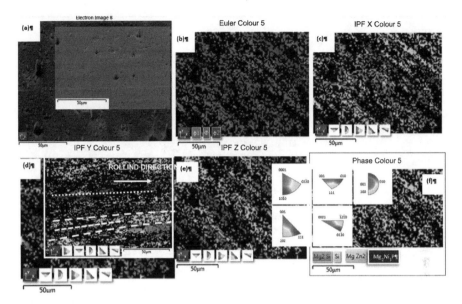

**FIG. 6.22**    EBSD analysis of the rolled 6016 Al at 20% thickness reduction

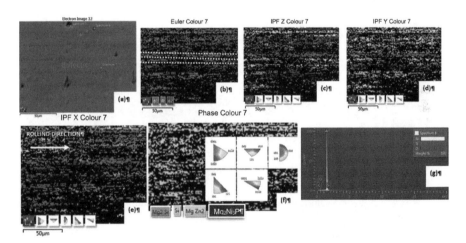

**FIG. 6.23**    EBSD analysis of the rolled 6016 Al at 60% thickness reduction

grains with a random orientation, alluding to texture configuration. Interestingly, the pole figures of the identified phases were established and were the same for the alloys (**Fig. 6.24**). Only the $\{110\}$ poles for the $Mg_2Ni_3P$ and $MgZn_2$, $\{100\}$ poles for $Mg_2Si$, and $\{111\}$ poles for eutectic-Si had the far-reaching intensity maxima's. The texture indices on the $\{110\}$ poles had maximum intensity of ~20 and 25 times random for the $Mg_2Ni_3P$ and $MgZn_2$ respectively; meanwhile, that of $Mg_2Si$ is ~60 times random; and lastly, the $\{111\}$ poles is ~20 times random.

**TABLE 6.4**
**Phase acquisition**

| Phase | a | b | c | Alpha | Beta | Gamma | Space group | Crystal structure |
|---|---|---|---|---|---|---|---|---|
| Mg2 Ni$_3$P | 4.97 Å | 4.97 Å | 10.96 Å | 90.00 ° | 90.00 ° | 120.00 ° | 166 | Rhombohedral |
| Mg$_2$ Si | 6.39 Å | 6.39 Å | 6.39 Å | 90.00 ° | 90.00 ° | 90.00 ° | 225 | BCC Cubic |
| Si | 5.41 Å | 5.41 Å | 5.41 Å | 90.00 ° | 90.00 ° | 90.00 ° | 227 | FCC Cubic |
| Mg Zn$_2$ | 5.23 Å | 5.23 Å | 8.57 Å | 90.00 ° | 90.00 ° | 120.00 ° | 194 | Tetragonal |

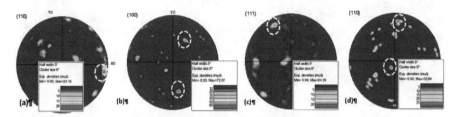

**FIG. 6.24**   Pole figures of the determined phases: (a) M2Ni3P, (b) Mg2Si, (c) eutectic-si, and (d) MgZn2

## Conclusions

The XRD and EBSD analysis of rolled 6016 Al alloy at 20% and 60% thickness reduction (trial runs) were investigated and the following outcomes were drawn:

- The XRD spectra indicated Al ss and eutectic-silicon.
- The W-H plots delineated that as the percentage reduction during rolling of 6016 Al alloy increases, the crystallite size increases, compared to its counterpart. This was associated with the accumulation of lattice strains in the grain boundaries.
- As expected, the longitudinal grains of the alloy thickness reduced at 60% is finer than its counterpart, as determined by the EBSD analysis.
- Eutectic-silicon was determined for both the alloys, and the remaining phases emanating from the Al ss were also designated in **Table 6.4** and confirmed by XRD peaks.
- Texture configuration was ascertained through the utility of pole figure maps. The most prominent texture indices were {110} poles for the Mg$_2$Ni$_3$P and MgZn$_2$, {100} poles for Mg$_2$Si, and {111} poles for eutectic-Si.

## Recommendations

- Optimizing EBSD parameters for indexing of Kikuchi patterns, considering the inherent phases formed either through processing or thermal treatment, as shown in **Fig. 6.25**.

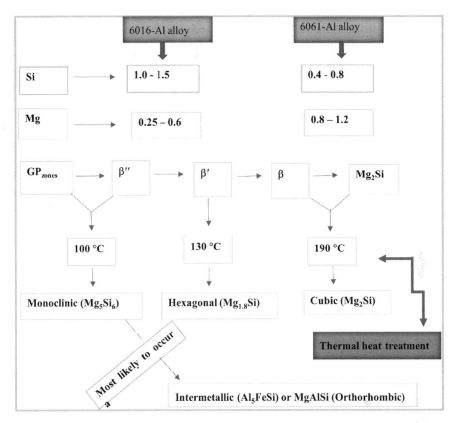

**FIG. 6.25** Schematic diagram illustration indicating compositional difference between the 6016 and 6061 Al alloys

- In order to access crystallographic data on the Twist software, computer simulation techniques using first-principles density functional theory will be used to simulate some of the possible phases incurred in the alloy, such as Al3Mg2 (Mg1.2Al1.8M2). In particular, the determination of atomic positions (Wyckoff notations), crystal system, Laue group and symbol, Schoenflies and their corresponding numbers, and lattice parameters will be carried out. An example modelled is shown in **Table 6.5** for an intermetallic (MgAlSi) shown in **Fig. 6.25**.

*MgAlSi orthorhombic*
- Lattice parameters (Å): a = 6.75; b = 4.05; c = 7.94
- Crystal system: PNMA
- Laue group: mmm
- Space group no: 62
- Symbol such as $P_2$ etc: Pearson symbol: oP12
- Alpha/beta/gamma angles: $\alpha = \beta = Y = 90$

- Schoenflies such as C1 or C2 and their corresponding numbers like 3b etc: D2H-16
- Atom occupancy numbers i.e. atomic coordinates using Wyckoff position for each element is shown in **Table 6.5**:

## Case study 2: Root-cause (failure) analysis of a broken boring mill shaft of material type SAE/AISI 4330 grade

Failure investigation was performed on a broken mill shaft by analyzing metallographic examination, chemical composition, impact test, hardness, and tempering heat-treatment processes, in order to locate the cause of failure. A transverse section of the shaft was removed from the initiation region of the fracture by cold cutting. This was placed in Bakelite and polished with polishing tools while being wet ground and diamond lapped to present a finish appropriate for microscopic scrutiny (**Fig. 6.26b**)

**TABLE 6.5**
**Wyckoff notation in the 6016 Al alloy**

| Site | Wyckoff position | Element | x | y | z | Occupancy |
|------|------------------|---------|--------|--------|--------|-----------|
| Mg1  | 4c               | Mg      | 0.03400 | 0.25000 | 0.67300 | 1.0 |
| Al1  | 4c               | Al      | 0.13900 | 0.25000 | 0.06800 | 1.0 |
| Si1  | 4c               | Si      | 0.26100 | 0.25000 | 0.38000 | 1.0 |

**FIG. 6.26**   The equipment utilized during the course of investigation: (a) ovens used for heat treatment, (b) polishing machines, (c) impact tester ISO 148, and (d) Vickers hardness tester

**TABLE 6.6**
**Chemical composition**

| Alloying elements | % Value |
|---|---|
| C | 0.357 |
| Si | 0.242 |
| Mn | 0.462 |
| P | 0.000 |
| S | 0.011 |
| Cu | 0.0180 |
| Al | 0.003 |
| Cr | 1.127 |
| Mo | 0.0263 |
| Ni | 1.532 |
| V | 0.012 |
| Ti | 0.004 |
| Nb | 0.003 |
| Co | 0.001 |
| W | 0.000 |
| Fe | 95.80 |

while other sections of the shaft was submitted for spectrometric analysis to ascertain chemical composition (**Table 6.6**) of the boring mill shaft.

A series of tempering temperatures were employed to determine whether there was any tempering embrittlement effect on the shaft material by measuring the depth of hardenability. The furnaces used for heat treatment are shown in **Fig. 6.26a** and were performed as follows:

- The sample was cut into four test pieces and on the first specimen core hardness was measured as received.
- The other test pieces were measured at varying temperatures of 250 °C, 350 °C, and 450 °C.

The impact strength was carried out as per the ISO 148 standard using an impact tester (**Fig. 6.26c**) while a Vickers hardness traverse test (**Fig. 6.26d**) in accordance with BS EN ISO 6507-1: 2005 was performed, using a 10kg load, from the surface of ground spline to the core area.

The general appearance of the shaft supplied is illustrated in **Fig. 6.27** below, where it can be seen that fracture had occurred at the change in section (circumferential groove/end of spline). **Fig. 6.28** shows the fracture face, characterized by herringbone or chevron patterns leading to the general area of the fracture origin, circled. The enlarged detail of this area is shown in **Fig. 6.29**, and shows probable two-fold origin areas, one at the undercut at the base of spline (A), and the other at the base of the circumferential groove (B). Both areas are black, unground surfaces (as heat treated), whereas the top and sides of the splines appear to be ground. Close examination of

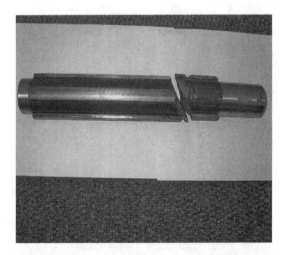

**FIG. 6.27**   As-received, visual examination of the shaft that had fractured in service

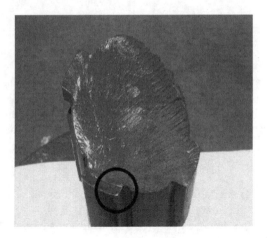

**FIG. 6.28**   Visual examination of the shaft that had fractured in service sectioned for fractography analysis

B shows that initial cracking was by fatigue at the section change at the base of the circumferential groove. The fatigued area is about 1mm deep, and fine ratchet marks were present, indicating reverse rotational service.

This composition is compared to the AISI 4330, 17CrNiMo6, and BS970 817M40 grades. The 17CrNiMo6 grade, having the lower carbon content, could be classed as a case-hardening steel, as opposed to the 817M40 (EN24) and AISI 4330 steel grades, which are used for through-hardening applications. It is important to note that the steels are Ni/Cr/Mo steels, with almost identical hardenabilities.

Examination (in the longitudinal plane) of the prepared section in the polished condition shows that the ASTM E45 cleanliness level is below 1.0 for all categories.

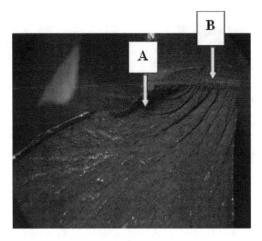

**FIG. 6.29**  Visual examination of the shaft that had fractured in service points showing crack initiation

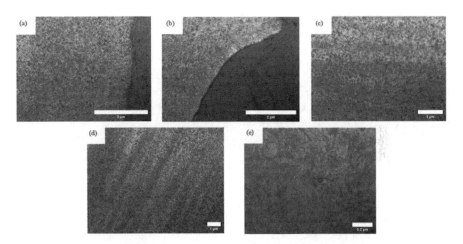

**FIG. 6.30**  Metallographic examination in the longitudinal plane: (a) at the ground surface, (b) near the cracked surface, (c) core at 100x, (d) core at 50x, and (e) core at 400x

In the 5% Nital etched condition, the structures are revealed in **Fig. 6.30a-d**. The core and near-surface structures consist of fine, untempered/lowly tempered martensite, having an ASTM E112 grain size of approx. 7.5. No evidence of a case-hardening layer was observed on the ground surface, **Fig. 6.30a**. The core matrix, **Fig. 6.30c–d**, was seen to exhibit moderate levels of elemental segregation/banding, which becomes prominent about 7mm below the surface. **Fig. 6.30b** shows decarburization near the crack initiation face, and this is consistent with the arrowed areas in **Fig. 6.29** (heat-treated, ungrounded surface in the grooves). This carbon-depleted surface is mainly untransformed ferrite, leaving a soft skin, as shown by the hardness traverse results

**TABLE 6.7**
**Variation of hardness from surface to the core (HV10)**

| | |
|---|---|
| At ground surface * | 572 |
| 0.3mm below the surface | 525 |
| 1.0 mm below the surface | 542 |
| 2.0 mm below the surface | 548 |
| 4.0mm below the surface | 548 |
| 6.0mm below the surface | 548 |
| 12.0mm below the surface | 536 |
| 14.0mm below the surface | 519 |
| 20.0mm below the surface | 525 |

*NB: The decarburized surface hardness (**Fig. 6.30b**) was found to be 290 HV10.

in **Table 6.7**. Lastly, **Fig. 6.30e** shows the core structure at higher magnification of untempered/lowly tempered martensite, which is expected to be brittle in nature, and could be susceptible to failure.

As the fracture appeared to be brittle, it was deemed necessary to measure the charpy impact strength, and the result was found to be 8J average. For a through-hardened steel which had been ideally hardened and sufficiently tempered, one would expect at least 35J. When gears and shafts are mainly designed to meet high toughness demands combined with increased fatigue properties, generally, tough-tempered medium-carbon steels are used. Typical mechanical properties are: hardness – 300 HB, yield Strength – 800 MPa, tensile Strength – 1000 MPa and impact 30 at – 20 °C [5]. The hardness traverse values are shown in **Table 6.7**. As it is expected that the hardness from the surface is low due to decarburization and measuring hardness toward the core, the values don't vary that much and this shows that the core of the shaft was not tough enough to withstand shock and impact during service (i.e. was brittle as a result of high hardness). This is indicative that the steel was not subjected to a high enough tempering process that would have aided in the toughening of the core.

**Fig. 6.31** demonstrates how hardness varies with increasing tempering temperatures. When the tempering temperature is increased, the hardness/strength level is decreased, while the degradation of impact strength is greatly minimized which then renders the steel to be tougher, possessing longer shelf life, including minimizing possible temper embrittlement.

The following conclusions were drawn:

1. Generally, fractures of shafts originate at points of stress concentration either inherent in design or introduced during fabrication or operation. In this case, where the base of the spline meets the circumferential groove (**Fig. 6.28–6.29**), there is an inherent stress, coupled with a sharp hardness gradient between the

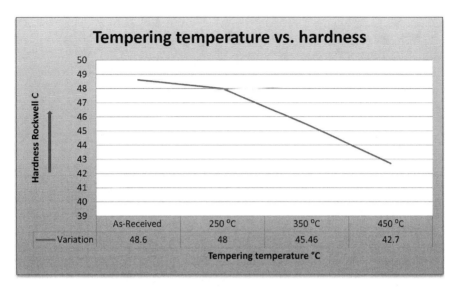

**FIG. 6.31**   The variation of tempering temperature with hardness. *NB: As hardness values decreases the impact strength values increases

outer skin and core, which will be magnified by the service stresses. Close examination of the fracture face near the origin shows that rotation/reverse rotation fatigue was involved as an initial crack formation, with sudden brittle fracture following shortly afterwards.

2.  However, the above features may not have solely contributed to the failure had the steel been heat treated correctly for the specified grade. For the steel in question, and bearing in mind the presence of chemical segregation, it would be advisable to carry out a high-temperature normalize, followed by an oil quench and temper to give the required hardness (the 'high' hardness of this shaft (535 HV ave., or 509 HB) indicates an almost as-quenched condition for this material grade and section, with a possibly very low tempering temperature carried out. This is also confirmed by the light-coloured etched microstructure (**Fig. 6.30**), and the low toughness, as seen from the impact test.

3.  The examination conducted was considered to reflect that the fracture experienced was directly related to the metallurgical condition of the shaft, which had imparted brittle characteristics to the material (high hardness and low toughness). Also, the stress concentration points coupled with possible impact service conditions (sudden stop/start) are considered contributing factors.

The following recommendations apply:

•  The microstructural condition of the steel indicates that it was through hardened (fully transformed to martensite) but lacked adequate tempering. However, for this a compromise between required hardness and toughness must be reached.

For higher hardness, tempering in the range 180–250 °C is recommended, and will not reduce the hardness significantly from what was measured for this shaft, but toughness will be reduced. However, for greater toughness, the range 460–630 °C is recommended, but not in the range 260–450 °C where temper embrittlement, or loss of notch toughness, may occur.

- For a diameter of this size it is expected that tempering at 580–620 °C will yield hardness similar to a condition 'T' or 'U' (250–330 HB). For the expected application it is considered that this strength is sufficient, yet giving high toughness, able to withstand impact loading and the effect of grooves or undercuts, if they are unavoidable. It has been found that components with these features can be 'surface toughened' by a shot peening operation. The client confirmed that wear resistance is much less of a factor than toughness in this application.

- The use of an alternative through-hardening steel is highly recommended, e.g. the BS970 817M40, 826M40 (Cond. T or U) or even the higher alloyed 835M30, where a higher hardness (if required), combined with good toughness, is achievable. However, the foregoing precautions of part geometry, heat treatment/tempering (for whatever grade and hardness level selected), etc., are to be considered when selecting a suitable material grade as suggested above.

## 6.5    MECHANICAL TESTING AND PROPERTIES

Mechanical properties of most materials are a more prevalent principal phenomenon with respect to contributing stability under force. For example, the deformation and fracture of materials depend on their structure when applied forces are used. When under intense forces, the macroscopic concerted effort of materials can disintegrate or even change shape. These changes in materials are scrutinized, examined experimentally, and identified with reference to the acting force per unit area viz., stress and displacement per unit strain [82]. However, in a perfect material such as whiskers without lattice flaws, modifications require extremely strong forces. Crystalline materials comprise a variety of lattice defects such as dislocations which provide simplicity of deformation. The materials under investigation are presumed to be homogenous vast continuums containing a variety of molecules, with negligible dynamic mechanisms between them similar to Ref. [82].

Mechanical properties influence mechanical ability and strength of material to be moulded in a desired shape, and can be defined as (1) strength – stems from stress-strain curve used to attain the materials Youngs modulus by correlating stress-strain values up to elastic limit; (2) stiffness – the ability of material to endure deformation under stress; (3) elasticity – property of material to recover its original shape after deformation provided external forces are released; (4) plasticity – when a material experiences deformation to a certain extent without failure [83]; (5) ductility – the material's ability to facilitate wire stretching during applied tensile force; (6) brittleness –the breaking of material with resultant damage; (7) malleability – the ability of material to be flattened, hammered, and rolled into thin sheets without damage during deformation; (8) toughness – the material's ability to resist bending without failure as a result of applied impact loads. A notched sample is utilized to ascertain notch sensitivity and impact energy, and some of the widely used standards include ASTM D256, ASTM E23, and ASTM D6110; (9) resilience – the material's ability to absorb energy and resist impact load and shock; (10) creep – a slow and permanent deformation when a material is subjected to constant stress at an elevated temperature for very long time; (11) fatigue – this is a tendency of materials to fail when subjected to cyclic loading, and is subdivided into three processes, including (a) first stage that induces crack nucleation and crack initiation, (b) crack propagation stage until uncracked cross-section cannot resist applied loads, and (c) fracture stage.[84]. It is important to note the deformations associated with low-cycle fatigue (LCF) and high-cycle fatigue (HCF), viz. LCF is represented by repeated deformation (in each cycle), while HCF is represented by elastic deformation. For example, both the tests can be performed according to the recommendation of ASTM E606 M12 specimen's holder on a 50kN Instron servo hydraulic testing machine [85] shown in **Fig. 6.32**; and (12) hardness – a material's ability to resist penetration by another material. There are various hardness measurements including, Vicker's hardness, Moh's scale, Knoop test, Brinell hardness (BHN), and Rockwell hardness, which measure area or depth of an indentation as a result of a specific applied force at a specific time [86]. ASTM standards for hardness tests are Brinell (ASTM E10), Knoop (ASTM E92, ASTM E384), and Vickers (ASTM E92, ASTM E384). For example, the Brinell hardness calculation is given in **Eqs. 6.1, 6.2**

$$BHN = \frac{Load\ on\ ball}{Area\ of\ indentation} \tag{6.1}$$

**FIG. 6.32** Instron model 1342

$$BHN = \frac{2P}{nD\left(D-\left(D^2-d^2\right)^{0.5}\right)} \tag{6.2}$$

where, $P$- is the applied load (kgf); $D$- is the diameter of the indentor ball (mm); $d$- is the diameter of the impression (mm) [84].

The ultimate failure strain of MMAMed alloys is lower than the conventional 316L alloy. This is contradictory, because for the same material, a specimen with a shorter gauge length exhibits a higher failure strain than a sample with a longer gauge length. This efficacity is regarded such that ductile materials elongate near the necking zone subsequent to reaching their tensile strength, but the measurable dimensions of the necking region are likely similar despite the length of the test specimen [87].

### 6.5.1  SPARK PLASMA SINTERING OF SHAPE MEMORY-BASED ALLOY: MECHANICAL CHARACTERISTICS AND MICROSTRUCTURAL DEVELOPMENT

### Case study 3: Microstructure evolution of Ti-Pt50-xMx ternary alloy during spark plasma sintering

*Abstract*

The purpose of the study involved solid-state sintering of elemental metallic powder blends using spark plasma sintering (SPS). The microstructural features and mechanical properties of Ti-Pt$_{50-x}$-M$_x$ (x = 25 at.%) and (M = Cu, Mn) ternary alloys were investigated by XRD, DSC, SEM, and microindentation hardness profile. The results indicate that during the microhardness determination, the bulk hardness decreases, and that transformation temperature shifted towards lower temperatures when adding a ternary element into the Ti-Pt binary system. Furthermore, the results also showed that SPS alone does not yield a homogeneous microstructure, thus annealing heat treatment was conducted for 24hrs to allow inter-diffusion of atoms.

*Background information*

Ti-Pt binary alloys have a thermo-elastic, martensitic, and displacive transformation from B2 (parent phase) to B19 (daughter phase), and exhibit properties such as shape memory effect & super elasticity [88]. The shape memory effect as stated by Yamabe-Mitarai et al. [89] occurs by reverse martensitic transformation, where the transformation temperature governs the operation temperature of shape memory alloys, and the authors further reported that in order to develop high-temperature shape memory alloys, the elevated martensitic transformation temperature is of importance.

Biggs et al. [90] described two types of phase transformations in the solid state of the Ti-Pt binary system. In the first, atoms fuse and deposition themselves into a new phase, each individual atom moving randomly. In the other type of transformation, whole rows of atoms shear or displace together, each atom moving in the same manner as its neighbours, to produce a significant increment of the new phase. The latter type of solid state phase transformation is known as displacive. The authors further reported that the Ti-Pt phase undergoes a reversible B2 ↔ B19 martensite transformation at ~1000 °C and provides the potential for shape memory effect (SME) at this temperature.

Furthermore, Lütjering et al. [91] reported that the martensitic transformation is the process that involves a co-operative movement of atoms resulting in a microscopically homogeneous transformation of bcc into hcp crystal structure. The following displacive transformations regarding Pt alloys are briefly described:

*Pt-Cr*

Preussner et al. [92] observed twinned microstructures in the 50 at%. Pt-Cr system which indicated that displacive transformation might have occurred.

*Ptzr*  Platinum group metals (PGMs) and their alloys have attracted considerable attention due to excellent physical, chemical, and mechanical properties; among these is the melting point of $Pt_3Zr$ with cubic $L1_2$ structure (2250 °C) and $Pt_4Zr$ (1880 °C) which is much higher than that of Ni-based super-alloys. In addition, PtZr alloy undergoes a martensite transformation from an ordered cubic structure to an orthorhombic structure, which is an indication of shape memory property [93].

Wadood et al. [94] argued that Zr belongs to the same group as titanium (group IV) in the periodic table, and the number of valence electrons for each group IV transition metal element is the same (four valence electrons). However, the atomic radius of Zr (0.155 nm) is larger than that of Ti (0.140 nm), so solid solution strengthening is expected due to the generation of compressive stresses by the partial substitution of Ti with Zr atoms having large atomic radii. The authors found out that the strength and shape memory effect (SME) of Ti-50Pt improved at high temperatures over 727 °C by partial (5 at.%) substitution of Ti with group IV transition metal Zr that exhibited an increase in critical stress for slip deformation related to the solid solution strengthening effect.

*Pt-Cu*  Biggs et al. [90] pointed out that the shortcomings of most commercial shape memory alloys (SMAs) is that neither of the common systems (NiTi or Cu-based) is able to function as SMAs at elevated temperatures. Moreover, Cu-based systems have an intrinsic problem with ageing, and undergo an associated time-dependent change in properties. Therefore, they must be protected from exposure to temperatures above 150 °C.

*Pt-Mn*  Pt-Cu and Pt-Mn illustrated twins when alloy screening was conducted [90]. In this study the transformation properties of Ti-Pt-Mn would be investigated if twinning microstructures are obtained since Mn is a bcc crystal structure different from Ti (hcp) & Pt (fcc).

The relevant properties of shape memory alloy (SMA) are its shape memory effect and pseudo elastic effect (PE). However, Fukuta et al. [95] described the shape memory effect (SME) of a wire for example; that when stretched at apparent yield point, residual strain will remain but when heating the wire to a particular temperature the residual strains disappears. The authors further pointed out that when the wire is heated to a different particular temperature it causes the strains to fully recover with unloading. According to Jani et al. [96], the significance of shape memory materials (SMMs) was not realized until the shape memory effect (SME) in a nickel-titanium alloy known as nitinol was discovered. However, the limitation of nitinol is its lower transformation temperature which prompted the demand in SMAs for engineering and technical high-temperature applications.

Inamura et al. [88] investigated the effects of ternary additions on Ms (martensite start) of TiNi and found out that most of the ternary elements such as Co, Fe, Mn, Cr, and V decrease Ms of TiNi whereas on the other hand, Hf, Zr and PGMs such as Au, Pd, and Pt are known to raise Ms of TiNi. Especially, the Pt addition to TiNi effectively raises the Ms temperature because the Ti-Pt binary alloy has a B2–B19 martensitic transformation of around 1027 °C and exhibits a shape memory effect. Furthermore, Biggs et al. [97] pointed out that this temperature can be readily varied even to as low as room temperature by partially substituting Pt with a third element such as nickel, and will exhibit transformation temperatures from ambient to 1000 °C. The present study will focus on partially substituting platinum with ternary elements, viz. copper and manganese, in order to observe whether HTSME is achievable in these alloys.

A number of binary intermetallic alloys with a B2 bcc structure at high temperatures transform displacively and reversibly to a B19 closer-packed martensitic structure at lower temperatures. Examples include: NiTi, NiAl, PdTi, AuTi and Ti-Pt. Of these, NiTi alloys are commercially exploited for use in shape memory applications and the $T_m$ (average $M_s$ and $A_s$ temperatures) lie between -100 °C and +100 °C. The $T_m$ temperatures of AuTi and PdTi are in the range of 400–600 °C. However, the $T_m$ temperature of Ti-Pt alloys is higher at around 1000 °C and this is considered to be of potential use for high-temperature shape memory alloys [97].

Wadood et al. [94] reported that equi-atomic Ti-Pt-based alloys could be used for higher-temperature shape memory material applications provided that their shape memory properties can be enhanced by increasing the critical stress for slip deformation at elevated temperatures. However, equi-atomic Ti-Pt alloys have low strength when mechanically tested at high temperatures (above 727 °C). Furthermore, increasing the critical stress for slip deformation will enhance the shape memory properties as well as the low-cycle fatigue properties of Ti-Pt alloys. However, increasing the stress for slip deformation will also improve the strength as well as high-cycle fatigue properties of Ti-Pt high-temperature shape memory alloys.

In this study, we also investigated the effect of partial substitution (25 at.%) of Pt with Cu and that of (25 at.%) Ti with Mn. The mechanical properties, shape memory effect, and phase transformation temperatures of Ti-50Pt alloys are of paramount importance for elevated-temperature applications. Furthermore, Wadood et al. [94] reported that for shape memory alloys there is a relationship between critical stress for slip deformation, critical stress for martensitic transformation/martensite re-orientation, and the shape memory effect. In addition the authors stated that for good shape memory effect the critical stress for slip deformation should be higher than the stress required for martensitic transformation/martensite re-orientation; else shape recovery will be adversely affected as more irreversible deformation will be induced by dislocation slip. However, the critical stress for slip deformation decreases with increasing temperature and if it becomes lower than the stress required for martensitic transformation/martensite re-orientation, the shape memory properties will decrease due to the occurrence of plastic deformation. Negligible shape memory effect and single yielding of the Ti–50Pt alloy at 910 °C below the martensite finish temperature are related to low critical stress for slip deformation compared to the stress required for martensite re-orientation of B19 orthorhombic martensite, as only the B19 martensitic phase is observed at room temperature. The main aim of this study was to investigate if we can improve the high-temperature, mechanical, and shape memory properties of the Ti–50Pt alloy by increasing the critical stress for slip deformation due to partial substitution of Ti and Pt with Mn and Cu, respectively.

*Objectives*

- To determine Ti-Pt$_{50-x}$-M$_x$ using spark plasma sintering (SPS) of blended elemental (BE) Ti-Pt-Cu, and Ti-Pt-Mn ternary powders.
- To optimize the SPS parameters that would give the starting composition.
- To produce a homogeneous Ti-Pt$_{50-x}$-M$_x$ alloy phase.
- To investigate the microstructural features and mechanical properties of Ti-Pt$_{50-x}$-M$_x$ (x = 25 at.%) and (M = Cu, Mn) ternary alloys employing XRD, DSC, SEM, and microindentation hardness profile.
- According to Inamura et al. [88], the martensitic transformation temperature (Ms) controls the actuation temperature of shape memory alloys (SMA). To determine which ternary element has the least or greatest impact on raising the shape memory temperature of the binary alloy, one goal of the study is to calculate the amount by which the SME temperature decreases when different compositions of ternary Cu and Mn are added to the Ti-Pt system.
- To establish a ternary alloy that has a martensitic transformation temperature above 100 °C. The target temperature for this work will be in the range of 300–700 °C.

The study is important as it expands literature on HTSMAs, and a new alloy will be recommended for application in high-temperature environments.

*Materials and methods*

The powder used in this study was the CSIR Ti. The powder particle size was quoted as −325 mesh (-45 μm): powder particle size analysis using a Microtrac® Bluewave laser particle size analyzer showed the D50 size was 31.25 μm (**Fig. 6.33a, b**). The Cu powder showed D10 = 15μm, D50 = 21.26μm, and D90 = 45μm with the irregular particle shape shown in **Fig. 6.34 a, b**. The Mn powder showed D10 = 21μm, D50 = 31.33μm, and D90 = 61μm, also with an irregular shape (**Fig. 6.35a, b**). Lastly the morphology of Pt powder is shown in **Fig. 6.36**.

The SEM image **Fig. 6.33** shows x1000 of elemental Ti powder which has been gas atomized showing nearly perfect spheres with globular-shaped particles.

The SEM x1000 image **Fig. 6.34** shows the water atomized elemental Cu powder. The arrow D shows where the bond has broken whereas the arrow S outlines the smooth, rounded surface of a particle that did not bond to the adjacent particle above

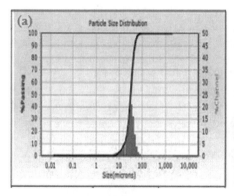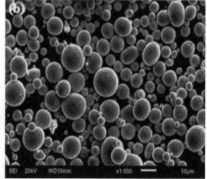

**FIG. 6.33**   CSIR Ti powder (a) particle size distribution, and (b) SEM electron micrograph

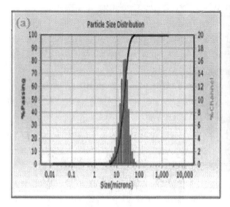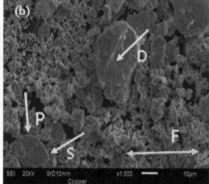

**FIG. 6.34**   Cu powder (a) particle size distribution, and (b) SEM electron micrograph

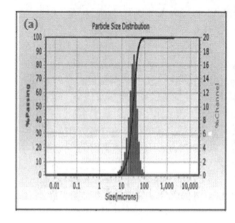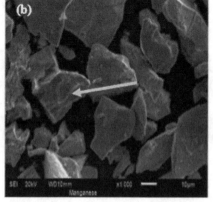

**FIG. 6.35**   Mn powder (a) particle size distribution, and (b) SEM electron micrograph

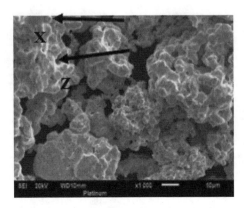

**FIG. 6.36**   Pt powder SEM electron micrograph

it. The double arrow F indicates the fibrous-like morphology illustrating the degree of irregularity and roughness. Arrow P shows particle boundaries that will disappear during proper sintering. The shape of the Cu particles is fibrous and irregular.

The SEM image at x1000 **Fig. 6.35** is an elemental Mn powder x1000 indicating coarse angular-like shape. The arrow indicates where the bond may have broken during the powder atomization process. The shape of the Mn particles assumes an angular (polyhedral) shape; this morphology shows many faces that act as a base.

The SEM image at x1000 **Fig. 6.36** is a water atomized elemental Pt powder. Arrow X indicates small fines that were agglomerated onto larger particles. Arrow Z shows porosity in spongy regions. The shape of the Pt particles assumes a spongy-like morphology with a high surface area to volume ratio.

The alloy composition has been calculated to convert the individual elemental powder from atomic percentage to weight in grams. This was to determine how much of each composition is needed based on the availability of metallic powder. The starting alloy was assumed to be 6g of each respective composition. After charge calculation, characterization of each elemental powder was performed. Furthermore, the Ti–25Pt–25Cu (at.%) and Ti–25Pt–25Mn (at.%) alloy powders were prepared by blending the composition in a tubular mixer for 1 hour; thereafter it was analyzed for SEM. Thereafter, sintering of the alloy powders using a tube argon flushed furnace **Fig. 6.37** was performed using the following parameters:

Ti-25Pt-25Cu = 900 °C; 20 minutes
Ti-25Pt-25Mn = 1100 °C; 20 minutes

Consolidation of the powders was followed by powder characterization viz: SEM, XRD, microhardness. Lastly post-heat treatment was done to homogenize the sintered alloy composition. The heat-treatment process was done in a tubular furnace for 24 hours, then cooled in a furnace; thereafter SEM, DSC, and microhardness was performed.

**FIG. 6.37**  Tube furnace

## Results and discussion
*XRD analysis*
*Ti elemental powder (***Fig. 6.38***)*

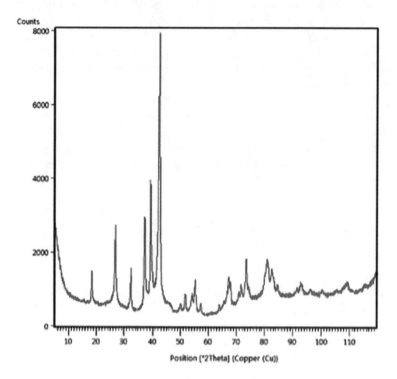

**FIG. 6.38**  XRD Results for Ti powder with the highest peak at 41 ° 2theta angle whereas that of Pt highest peak is 35 ° 2theta angle

*Cu elemental powder (***Fig. 6.39***)*

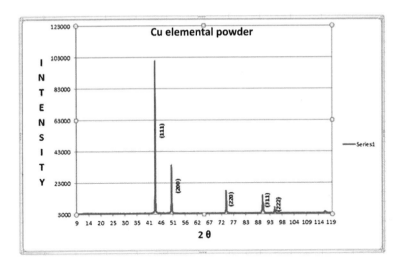

**FIG. 6.39** XRD of Cu elemental powder indicating the bulk composition of the powder as copper and the remainder 1% the residual Al & O. The close packed plane {111} having the highest intensity at 43 ° 2theta angle

*Mn elemental powder (***Fig. 6.40***)*

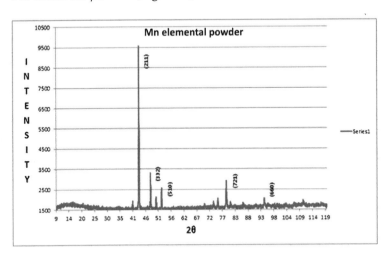

**FIG. 6.40** XRD Mn elemental powder indicating the bulk composition of the powder as Mn (96%) and the remainder 4% (the residual Zn & O) according to the spectrum analysis of the EDS. The close packed plane {111} having the highest intensity at 43 ° 2theta angle

However, the XRD results for blended alloy composition, viz: Ti-25Pt-25Cu and Ti-25Pt-25Mn showed the same diffraction patterns as that of individual elemental powders as indicated in **Figs. 6.41** and **6.42**.

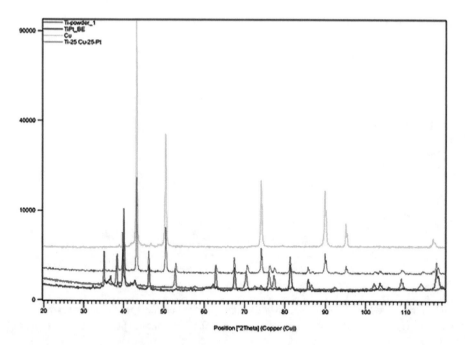

**FIG. 6.41**   Ti-25Pt-25Cu blended powder

**Fig.6.41** shows the XRD of Ti-25Pt-25Cu blended elemental (BE) powder indicating the highest peak (Cu) at 43 ° 2theta angle with a (101) plane, Ti highest peak at 41 (002) plane and lastly the Pt highest peak at 35 ° 2theta angle.

**Fig. 6.42** shows X-ray diffraction of Ti-25Pt-25Mn blended elemental (BE) powder indicating the highest peak (Ti) at 29 ° 2theta angle with a (220) plane, Pt highest peak at 35 ° (222) plane, and lastly the Mn highest peak at 43 ° 2theta.

*Characterization – Post-sintered XRD results*

*Ti-25Pt-25Cu (***Fig. 6.43***)*   **Fig. 6.43** illustrates a peak at 24 ° 2theta angle which has not been identified in the diffraction pattern of single elemental powders.

*Ti-25Pt-25Mn (***Fig. 6.44***)*   **Fig. 6.44** shows peaks at 24 ° and 26 ° 2theta angle which has not been identified in the diffraction pattern of single elemental powders.

The results of XRD pattern of Ti-25Pt-25Cu & Ti-25Pt-25Mn blended elemental powder alloy composition indicate that the sintering process was successful but doesn't indicate whether the compositions were homogeneous. However, the SEM was able to confirm homogeneity to a certain extent which is discussed in the following section.

*SEM results*

*Ti-25Pt-25Cu as-sintered*   The following shows the SEM micrograph, each phase showing chemical composition obtained via EDS.

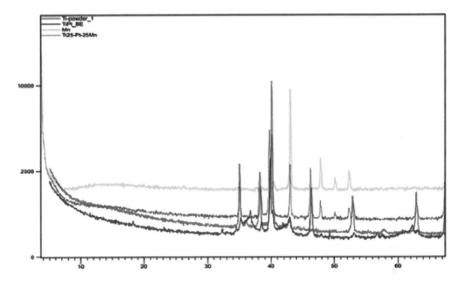

**FIG. 6.42**   Ti-25Pt-25Mn blended powder

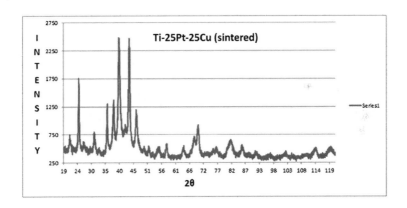

**FIG. 6.43**   XRD results for Ti-25Pt-25Cu sintered powder with the highest peak at 41 ° & 43 ° 2theta angles

The chemical composition of the individual phases in **Fig. 6.45** is illustrated by **Tables 6.8–6.11**. The white phase shows the platinum rich area indicated by **Table 6.8**, the titanium and copper rich phase is indicated in **Table 6.9**, the phase shown by the red arrow is indicated in **Table 6.11**, whereas the chemical composition shown by the green arrow is in **Table 6.10**.

The SEM micrograph shows that the ternary alloy is not homogeneous, since it's indicated by EDS results demonstrating different phases throughout the microstructure. However, the chemical composition in **Table 6.12** shown by the orange double arrow (**Fig. 6.45**) might indicate the starting ternary alloy composition since Ti is

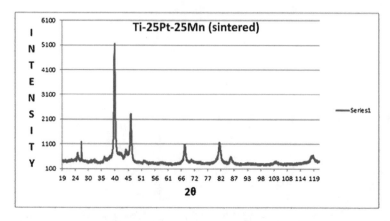

**FIG. 6.44**  XRD Results for Ti-25Pt-25Mn sintered powder with the highest peak at 40 ° 2theta angles

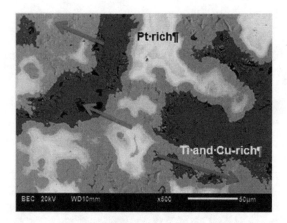

**FIG. 6.45**  SEM micrograph of Ti-25Pt-25Cu depicted for analysis of the phase determination using EDS

**TABLE 6.8**
**Platinum rich**

| Element | at.% | wt.% |
|---------|------|------|
| Ti      | 20.1 | 6.1  |
| Pt      | 73.3 | 91.2 |
| Cu      | 6.6  | 2.7  |

**TABLE 6.9**
**Ti and Cu rich**

| Element | at.% | wt.% |
|---------|------|------|
| Ti | 70.4 | 61.0 |
| Pt | 2.1 | 7.4 |
| Cu | 27.5 | 31.6 |

**TABLE 6.10**
**Chemical composition of the alloy**

| Element | at.% | wt.% |
|---------|------|------|
| Ti | 50.0 | 23.9 |
| Pt | 33.8 | 65.8 |
| Cu | 16.2 | 10.3 |

**TABLE 6.11**
**Ti rich area**

| Element | at.% | wt.% |
|---------|------|------|
| Ti | 97.5 | 94.9 |
| Pt | 1.0 | 3.8 |
| Cu | 0.9 | 1.2 |

**TABLE 6.12**
**Ternary alloy composition**

| Element | at.% | wt.% |
|---------|------|------|
| Ti | 50.0 | 23.9 |
| Pt | 33.8 | 65.8 |
| Cu | 16.2 | 10.3 |

(50 at.%). Furthermore, annealing heat treatment was performed to homogenize the microstructure, and the alloys were annealed in a tube furnace at 1100 °C soaked for 24 hours before being cooled in the furnace. The scatter of different phases was minimal but rather that the ternary composition was found to be more profound and uniformly distributed throughout the microstructure after homogenization anneal.

The chemical composition of a phase indicated by the blue arrow in **Fig. 6.46** is shown in **Table 6.13**, whereas that of the purple arrow is shown in **Table 6.14**.

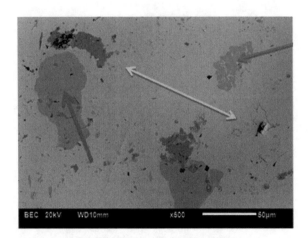

**FIG. 6.46**   SEM micrograph of Ti-25Pt-25Cu homogenized

**TABLE 6.13**
**Ti rich phase chemical composition**

| Element | at.% | wt.% |
|---|---|---|
| Ti | 50.0 | 23.9 |
| Pt | 33.8 | 65.8 |
| Cu | 16.2 | 10.3 |

**TABLE 6.14**
**Ti with Pt rich composition**

| Element | at.% | wt.% |
|---|---|---|
| Ti | 50.0 | 23.9 |
| Pt | 33.8 | 65.8 |
| Cu | 16.2 | 10.3 |

*Ti25Pt-25CMn as-sintered* (**Fig. 6.47**) **Fig. 6.47** shows the SEM micrograph of Ti-25Pt-25Mn. The microstructure shows scatters of different phases and thereafter indicates that there is inhomogeneity throughout the micrograph. The same principle was applied for homogenizing the structure; and the profile used was that the as-sintered alloy composition was annealed in a tube furnace at 1100 °C, soaked for 24 hours, followed by furnace cooling. Furthermore, the chemical composition of each phase obtained via EDS is designated by **Tables 6.15–6.18**. The white area is rich in platinum, shown in **Table 6.15**, the chemical composition of the blue arrow is Mn rich, shown in **Table 6.16**, the area shown by the red arrow is Ti and Mn rich, shown in **Table 6.17**, and lastly the green arrow is Ti rich, shown in **Table 6.18**.

**FIG. 6.47**   SEM micrograph of Ti-25Pt-25Mn depicted for analysis of the phase determination using EDS

**TABLE 6.15**
**Pt rich area**

| Element | at.% | wt.% |
|---------|------|------|
| Ti      | 11.9 | 3.3  |
| Pt      | 85.2 | 95.8 |
| Mn      | 2.9  | 0.9  |

**TABLE 6.16**
**Mn rich composition**

| Element | at.% | wt.% |
|---------|------|------|
| Ti      | 42.2 | 37.7 |
| Pt      | 1.2  | 4.4  |
| Mn      | 56.3 | 57.7 |

**TABLE 6.17**
**Ti and Mn rich**

| Element | at.% | wt.% |
|---------|------|------|
| Ti      | 55.8 | 46.9 |
| Pt      | 4.3  | 14.6 |
| Mn      | 39.9 | 38.5 |

**TABLE 6.18**
**Ti rich**

| Element | at.% | wt.% |
|---------|------|------|
| Ti | 64.3 | 36.7 |
| Pt | 24.0 | 55.7 |
| Mn | 11.7 | 7.6 |

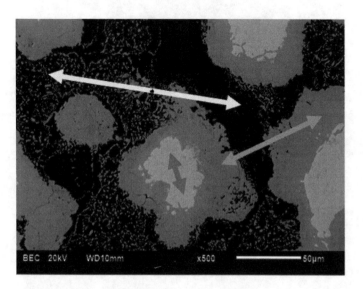

**FIG. 6.48**   SEM micrograph of Ti-25Pt-25Mn homogenized

**Fig. 6.48** below is an indication that the micrograph is somehow homogeneous when compared to **Fig. 6.47**. The orange double arrow in **Fig. 6.48** points to a composition in **Table 6.19** which is somehow closer to the starting ternary alloy Ti-25Pt-25Mn, because it may be that the B2–B19 transformation has its stability range in the Ti-Pt binary phase diagram extending at about 45–56 at.%Pt and 56–45 at.%Ti. The double arrow red indicates a phase that is slightly rich in platinum (**Table 6.20**). Lastly the composition in **Table 6.21** of **Fig. 6.48** is represented by the yellow arrow.

*Microhardness profile*

The microindentation hardness profile was performed after the annealing heat treatment. The traverse hardness values performed throughout the microstructure may have been designated to show homogeneity of the phases in the microstructure.

*Ti-25Pt-25Cu (**Fig. 6.49**)*   The following figure shows traverse hardness profile succeeding the post-annealing process indicated by both the scatter and bar graphs.

**TABLE 6.19**
**Ternary starting composition**

| Element | at.% | wt.% |
|---------|------|------|
| Ti | 55.9 | 31.8 |
| Pt | 23.7 | 54.9 |
| Mn | 20.4 | 13.3 |

**TABLE 6.20**
**Platinum rich phase**

| Element | at.% | wt.% |
|---------|------|------|
| STi | 30.3 | 12.4 |
| Pt | 46.1 | 11.0 |
| Mn | 23.6 | 76.6 |

**TABLE 6.21**
**Ti rich phase composition**

| Element | at.% | wt.% |
|---------|------|------|
| Ti | 61.8 | 51.6 |
| Pt | 4.9 | 16.8 |
| Mn | 32.2 | 30.9 |

**Fig. 6.49** (**A & B**) indicates that the microstructure shows more or less the same hardness values, with **Fig. 6.49B** showing the error bars overlapping which may be an indication that the results are statistically similar.

*Ti-25Pt-25Mn (***Fig. 6.50***)* **Fig. 6.50** illustrates the microhardness indentation profile post-annealing process for a Ti-25Pt-25Mn alloy. The homogeneity of the micrograph in **Fig. 6.48** above was shown through the hardness values in **Fig. 6.50** (**A & B**), with **Fig. 6.50B** with the error bars overlapping indicating that there was no significant difference and that the results are statistically similar.

*DSC results*

The following figures show DSC results for the Ti-25Pt-25Cu and Ti-25Pt-25Mn.

*Ti-25Pt-25Cu (***Fig. 6.51***)* **Fig. 6.51** demonstrates the intermetallic homogeneous phase structure formed during the phase transition starting at 1229 °C. During cooling the DSC results show another phase starting to form at 1268 °C. These results further

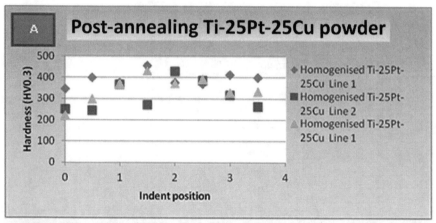

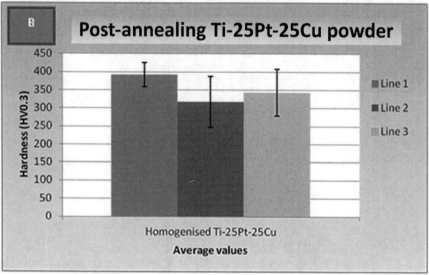

**FIG. 6.49**   (A & B): Hardness profile represented both in scatter and bar graphs for Ti-25Pt-25Cu alloy

show that the displacive transformation did not occur, in contrast to the Ti-Pt binary system; instead intermetallic phases were formed at these temperatures.

*Ti-25Pt-25Mn* (**Fig. 6.52**)   **Fig. 6.52** illustrates sharper peaks which might represents melting of the manganese element, which melts at about a temperature of (1246 °C) when compared to Ti and Pt, whereas the small troughs may be considered to show an unknown intermetallic homogeneous phase, of which during heating the onset temperature is 1165 °C and during cooling the onset temperature is 1226 °C. Furthermore, like **Fig. 6.51**, these results show that the displacive transformation did

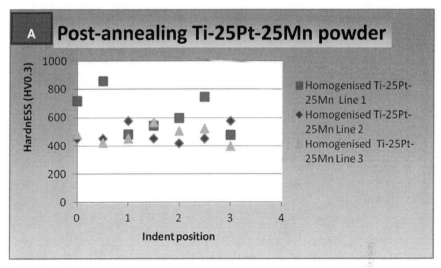

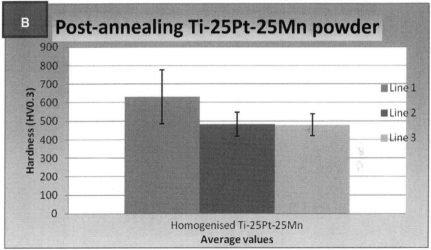

**FIG. 6.50** (A & B): Hardness profile represented both in scatter and bar graphs for Ti-25Pt-25Mn alloy

not take place as compared to the Ti-Pt binary system, but instead an intermetallic phase occurred at these temperatures.

*Conclusions*

- The results showed that SPS alone does not yield a homogeneous microstructure.
- Homogenization was achieved after annealing for 24 hours in a tube furnace; this was backed up by the microindentation hardness profile.

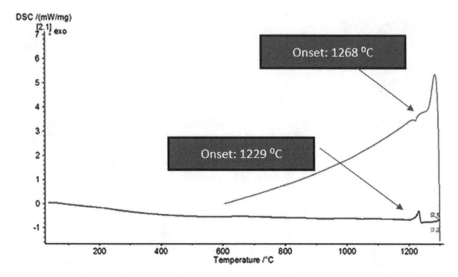

**FIG. 6.51**   DSC results for Ti-25Pt-25Cu

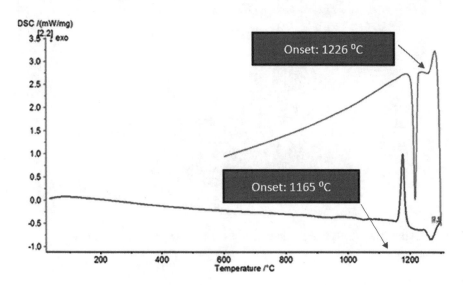

**FIG. 6.52**   DSC results for Ti-25Pt-25Mn

- Compared to Ti-Pt binary alloys, the addition of ternary alloying elements lowered the displacive transformation temperature.
- The DSC results didn't show any reversible martensitic transformation, instead homogeneous phases that occur at such transformation temperatures.
- Further work on structural analysis such as TEM would give a clear indication regarding the type of homogeneous phases.

## 6.6 CONCLUSIONS

The spectrum of material characterization has a vast number of capabilities and techniques that include imaging and microanalysis, and surface, structural, organic, elemental, thermal, and mechanical analysis. The advancement in material research is the development of methods and characterization of complex materials to perform extensive forensic studies and provide solutions. Moreover, this chapter has presented expertise that can trace aberrations and establish novel approaches when conventional techniques do not provide resolutions that are fundamental to unravel some of the world's inflexible problems.

## REFERENCES

[1] B. Rankouhi, Z. Islam, F. E. Pfefferkorn, & D. J. Thoma, Characterization of multi-material 316L-Hastelloy X fabricated via laser powder-bed fusion, *Mater. Sci. Eng. A,* 837, (January), 142749, 2022, doi: 10.1016/j.msea.2022.142749.

[2] A. K. Kercher, Statistical methods for material characterization and qualification, Oak Ridge National Lab. (ORNL), U.S. Department of Energy Office of Scientific and Technical Information, United States: N. p., 2005, doi:10.2172/886013.'

[3] D. J. Wulpi, *Failures of shafts in ASM handbook: Failure analysis and prevention,* ASM International, 2002.

[4] I. Kharagpur, 22 Shaft and its design based on strength, version 2 ME, 2014. [Online]. Available: http://nptel.ac.in/courses/112105125/pdf/mod8les1.pdf

[5] E. Claesson, Head of Industry Solution Development, Ovako Steel AB: Sweden, www.ovako.com.

[6] C. Nyberg, Understanding factors that cause shaft failures. *Pump & Services,* June, 2007.

[7] I. Rexnord, *Failure analysis of gears, shafts, bearings & seals,* Canal St., Milwaukee: Rexnord 1978.

[8] C. P. Vincze, J. S. Greenlaw, L. Ramey, & M. Faecher, *Investigating material and component failures: Structural engineering,* TRC Companies Inc.,2004.

[9] B. M. Gliner, *Determination of the Mechanical and Technological properties of Metals,* 2nd ed., New York, Oxford, London, Paris: Pergamon Press Ltd, 1960.

[10] V. Milman, S. J. Pennycook, & D. E. Jesson, Ab initio total energy study of adsorption and diffusion on the Si (100) surface, *Thin Solid Films,* 272, 375–385, 1996.

[11] C. Jiang, First-principles study of site occupancy of dilute 3d, 4d and 5d transition metal solutes in L10 TiAl, *Acta Mater.,* 56, (20), 6224–6231, 2008, doi: 10.1016/j.actamat.2008.08.047.

[12] M. N. Mathabathe, A. S. Bolokang, G. Govender, C. W. Siyasiya, & R. J. Mostert, Characterization of the nitrided γ-Ti-46Al–2Nb and γ-Ti-46Al–2Nb-0.7Cr-0.3Si intermetallic alloys, *Mater. Chem. Phys.,* 257, (December 2019), 123703, 2021, doi: 10.1016/j.matchemphys.2020.123703.

[13] M. N. Mathabathe, G. Govender, C. W. Siyasiya, R. J. Mostert, & A. S. Bolokang, Surface characterization of the cyclically oxidized γ-Ti-48Al-2Nb-0.7Cr alloy after nitridation, *Mater. Charact.,* 154, 94–102, 2019, doi: 10.1016/j.matchar.2019.05.036.

[14] S. R. Chubb & D. A. Papaconstantopoulos, First-principle study of L10 Ti-Al and V-Al alloys, *Phys. Rev. B,* 38, 12120, 1988.

[15] M. N. Mathabathe, R. Modiba, & A. S. Bolokang, The effects of quaternary alloying additions on the γ TiAl alloy: Preferential site occupancy, interfacial energetics to

physical parameters, *Surfaces and Interfaces*, 25, (May), 101173, 2021, doi: 10.1016/j.surfin.2021.101173.

[16]  S. I. Ranganathan & M. Ostojastarzewski, Universal elastic anisotropy index, *Phys. Rev. Lett,* 101, 055504, 2008.

[17]  J. Zhang, H. Li, X. Sun, & M. Zhan, A multi-scale MCCPFEM framework: Modeling of thermal interface grooving and deformation anisotropy of titanium alloy with lamellar colony, *Int. J. Plast.,* 135, (July), 102804, 2020, doi: 10.1016/j.ijplas.2020.102804.

[18]  O. Ouadah, G. Merad, F. Saidi, S. Mendi, & M. Dergal, Influence of alloying transition metals on structural, elastic, electronic and optical behaviors of γ-TiAl based alloys: A comparative DFT study combined with data mining technique, *Mater. Chem. Phys.,* 242, (August 2019), 122455, 2020, doi: 10.1016/j.matchemphys.2019.122455.

[19]  W. Voigt, *Leipzig: Lehrbuch der Kristallphysik.* Wiesbaden: Springer Fachmedien Wiesbaden GmbH, 1966. doi: DOI 10.1007/978-3-663-15884-4.

[20]  A. Reuss & Z. Angew, Calculation of the flow limit of mixed crystals on the basis of the plasticite at condition for single crystals, *Math. Mech*, 9, 49, 1929.

[21]  R. Hill, The elastic behaviour of a crystalline aggregate, *Proc. Soc. London*, 65, 439, 1952.

[22]  S. F. Pugh, Relations between the elastic moduli and the plastic properties of polycrystalline pure metals, *Philos. Mag*, 45, (367), 823–843, 1954.

[23]  C. Suryanarayana & M. G. Norton, *X-ray diffraction: A practical approach*, New York: Springer Science & Business Media, 2013.

[24]  M. Stefanou, K. Saita, D. V. Shalashilin, & A. Kirrander, Comparison of ultrafast electron and X-ray diffraction – A computational study, *Chem. Phys. Lett.*, 683, 300–305, 2017, doi: 10.1016/j.cplett.2017.03.007.

[25]  M. Hosseini, M. Arif, A. Keshavarz, & S. Iglauer, Neutron scattering: A subsurface application review, *Earth-Science Rev.*, 221, (June), 103755, 2021, doi: 10.1016/j.earscirev.2021.103755.

[26]  C. Mansas, C. Rey, X. Deschanels, & J. Causse, Scattering techniques to probe the templating effect in the synthesis of copper hexacyanoferrate nanoparticles via reverse microemulsions, *Colloids Surfaces A Physicochem. Eng. Asp.*, 624, (March), 2021, doi: 10.1016/j.colsurfa.2021.126772.

[27]  D. T. Pierce, D. R. Coughlin, D. L. Williamson, K. D. Clarke, A. J. Clarke, J. G. Speer, & E. De Moor, Characterization of transition carbides in quench and partitioned steel microstructures by Mössbauer spectroscopy and complementary techniques, *Acta Mater.*, 90, 417–430, 2015, doi: 10.1016/j.actamat.2015.01.024.

[28]  F. A. Selim, D. P. Wells, J. F. Harmon, J. Kwofie, G. Erikson, & T. Roney, New positron annihilation spectroscopy techniques for thick materials, *Radiat. Phys. Chem.*, 68, (3–4), 427–430, 2003, doi: 10.1016/S0969-806X(03)00249-4.

[29]  M. Fittipaldi, D. Gatteschi, & P. Fornasiero, The power of EPR techniques in revealing active sites in heterogeneous photocatalysis: The case of anion doped TiO2, *Catal. Today*, 206, 2–11, 2013, doi: 10.1016/j.cattod.2012.04.024.

[30]  Y. Zhu, X. Qiu, X. Chen, M. Huang, & Y. Li, Single gold nanowire-based nanosensor for adenosine triphosphate sensing by using in-situ surface-enhanced Raman scattering technique, *Talanta*, 249, (April), 123675, 2022, doi: 10.1016/j.talanta.2022.123675.

[31]  R. Cuscó, E. Alarcón-Lladó, J. Ibáñez, L. Artús, J. Jiménez, B. Wang, & M. J. Callahan, Temperature dependence of Raman scattering in ZnO, *Phys. Rev. B*, 75, (16), 165202, 2007, doi: 10.1103/PhysRevB.75.165202.

[32] M. N. Mathabathe, A. S. Bolokang, G. Govender, C. W. Siyasiya, & R. J. Mostert, Cold-pressing and vacuum arc melting of γ-TiAl based alloys, *Adv. Powder Technol.*, 30, (12), 2925–2939, 2019, doi: 10.1016/j.apt.2019.08.038.

[33] A. S. Bolokang, Z. P. Tshabalala, G. F. Malgas, I. Kortidis, H. C. Swart, & D. E. Motaung, Room temperature ferromagnetism and CH4 gas sensing of titanium oxynitride induced by milling and annealing, *Mater. Chem. Phys.*, 193, 512–523, 2017, doi: 10.1016/j.matchemphys.2017.03.012.

[34] A. S. Bolokang, D. E. Motaung, C. J. Arendse, & T. F. G. Muller, Formation of the metastable FCC phase by ball milling and annealing of titanium-stearic acid powder, *Adv. Powder Technol.*, 26, (2), 632–639, 2015, doi: 10.1016/j.apt.2015.01.013.

[35] A. S. Bolokang, D. E. Motaung, C. J. Arendse, & T. F. G. Muller, Morphology and structural development of reduced anatase-TiO2 by pure Ti powder upon annealing and nitridation: Synthesis of TiOx and TiOxNy powders, *Mater. Charact.*, 100, 41–49, 2015, doi: 10.1016/j.matchar.2014.11.026.

[36] M. Naguib, V. Presser, D. Tallman, J. Lu, L. Hultman, Y. Gogotsi, & M. W. Barsoum, On the topotactic transformation of Ti2AlC into a Ti-C-O-F cubic phase by heating in molten lithium fluoride in air, *J. Am. Ceram. Soc.*, 94, (12), 4556–4561, 2011, doi: 10.1111/j.1551-2916.2011.04896.x.

[37] Z. Xiao, X. Zhu, Z. Chu, W. Xu, Z. Wang, & B. Wu, Investigation of Ti2AlC formation mechanism through carbon and TiAl diffusional reaction, *J. Eur. Ceram. Soc.*, 38, (October 2017), 1246–1252, 2017, doi: 10.1016/j.jeurceramsoc.2017.10.039.

[38] H. Luo, Y. Xiang, Y. Zhao, Y. Li, & X. Pan, Nanoscale infrared, thermal and mechanical properties of aged microplastics revealed by an atomic force microscopy coupled with infrared spectroscopy (AFM-IR) technique, *Sci. Total Environ.*, 744, 2020, doi: 10.1016/j.scitotenv.2020.140944.

[39] Y. Xiao, Y. Cheng, P. He, X. Wu, & Z. Li, New insights into external layers of cyanobacteria and microalgae based on multiscale analysis of AFM force-distance curves, *Sci. Total Environ.*, 774, 145680, 2021, doi: 10.1016/j.scitotenv.2021.145680.

[40] S. Theiner, A. Schoeberl, A. Schweikert, B. K. Keppler, & G. Koellensperger, Mass spectrometry techniques for imaging and detection of metallodrugs, *Curr. Opin. Chem. Biol.*, 61, 123–134, 2021, doi: 10.1016/j.cbpa.2020.12.005.

[41] T. F. Silva, C. L. Rodrigues, N. Added, M. A. Rizzutto, M. H. Tabacniks, T. Höschen, U. von Toussaint, & M. Mayer, Self-consistent ion beam analysis: An approach by multi-objective optimization, *Nucl. Instruments Methods Phys. Res. Sect. B Beam Interact. with Mater. Atoms*, 506, (September), 32–40, 2021, doi: 10.1016/j.nimb.2021.09.007.

[42] S. Landsberger & R. Kapsimalis, Comparison of neutron activation analysis techniques for the determination of uranium concentrations in geological and environmental materials, *J. Environ. Radioact.*, 117, 41–44, 2013, doi: 10.1016/j.jenvrad.2011.08.014.

[43] P. Y. Hu, Y. H. Hsieh, J. C. Chen, & C. Y. Chang, Characteristics of manganese-coated sand using SEM and EDAX analysis, *J. Colloid Interface Sci.*, 272, (2), 308–313, 2004, doi: 10.1016/j.jcis.2003.12.058.

[44] M. N. Mathabathe, A. S. Bolokang, G. Govender, R. J. Mostert, & C. W. Siyasiya, Structure-property orientation relationship of a gamma/alpha2/Ti5Si3 in as-cast Ti-45Al-2Nb-0.7Cr-0.3Si intermetallic alloy, *J. Alloys Compd.*, 765, 690–699, 2018, doi: 10.1016/j.jallcom.2018.06.265.

[45] D. J. Dingley & V. Randle, Review Microtexture determination by electron backscatter diffraction, *J. Mater. Sci.*, 27, 4545–4566, 1992.

[46] F. Sun, Integrated TEM/transmission-EBSD for recrystallization analysis in nickel-based disc superalloy, *Prog. Nat. Sci. Mater. Int.*, 31, (1), 63–67, 2021, doi: 10.1016/j.pnsc.2020.11.003.

[47] T. N. Blanton, C. L. Barnes, & M. Lelental, The effect of X-ray penetration depth on structural characterization of multiphase Bi-Sr-Ca-Cu-O thin films by X-ray diffraction techniques, *Phys. C Supercond. its Appl.*, 173, (3–4), 152–158, 1991, doi: 10.1016/0921-4534(91)90362-3.

[48] M. Teusner, J. Mata, & N. Sharma, Small angle neutron scattering and its application in battery systems, *Curr. Opin. Electrochem.*, 34, 100990, 2022, doi: 10.1016/j.coelec.2022.100990.

[49] F. Yokaichiya, C. Schmidt, J. Storsberg, M. Kumpugdee Vollrath, D. R. de Araujo, B. Kent, D. Clemens, F. Wingert, & M. K. K. D. Franco, Effects of doxorubicin on the structural and morphological characterization of solid lipid nanoparticles (SLN) using small angle neutron scattering (SANS) and small angle X-ray scattering (SAXS), *Phys. B Condens. Matter,* 551, (August 2017), 191–196, 2018, doi: 10.1016/j.physb.2017.12.036.

[50] M. S. Gaafar, H. A. Afifi, & M. M. Mekawy, Structural studies of some phospho-borate glasses using ultrasonic pulse-echo technique, DSC and IR spectroscopy, *Phys. B Condens. Matter*, 404, (12–13), 1668–1673, 2009, doi: 10.1016/j.physb.2009.01.045.

[51] M. Jebli, J. Dhahri, H. Belmabrouk, A. Bajahzar, & M. L. Bouazizi, Investigation of the effect of NbCl5 dopant on dielectric properties and Raman spectroscopy of Ba0.97La0.02TiO3 polycrystalline ceramic by molten-salt technique, *J. Mol. Struct.*, 1264, 133253, 2022, doi: 10.1016/j.molstruc.2022.133253.

[52] K. S. De Assis, A. C. Rocha, I. C. P. Margarit-Mattos, F. A. S. Serra, & O. R. Mattos, Practical aspects on the use of on-site double loop electrochemical potentiodynamic reactivation technique (DL-EPR) for duplex stainless steel, *Corros. Sci.*, 74, 250–255, 2013, doi: 10.1016/j.corsci.2013.04.050.

[53] T. Alharbi, H. F. M. Mohamed, Y. B. Saddeek, A. Y. El-Haseib, & K. S. Shaaban, Study of the TiO2 effect on the heavy metals oxides borosilicate glasses structure using gamma-ray spectroscopy and positron annihilation technique, *Radiat. Phys. Chem.*, 164, (April), 108345, 2019, doi: 10.1016/j.radphyschem.2019.108345.

[54] H. Aziam, S. Indris, H. Ben Youcef, R. Witte, A. Sarapulova, H. Ehrenberg, & I. Saadoune, The first lithiation/delithiation mechanism of MFeOPO4 (M: Co, Ni) as revealed by 57Fe Mössbauer spectroscopy, *J. Alloys Compd.*, 906, 164373, 2022, doi: 10.1016/j.jallcom.2022.164373.

[55] T. Wang, M. Bi, J. Wu, X. Li, Y. Meng, Z. Yin, & W. Hang, Single-cell mass spectrometry imaging of TiO2 nanoparticles with subcellular resolution, *Chinese J. Anal. Chem.*, 50, (5), 100085, 2022, doi: 10.1016/j.cjac.2022.100085.

[56] C. Cheng, D. Hei, W. Jia, Q. Shan, Y. Ling, & C. Shi, Metallic materials analysis by DT neutron generator-based inelastic neutron scattering system: Measurement and Monte Carlo simulation, *Nucl. Instruments Methods Phys. Res. Sect. B Beam Interact. with Mater. Atoms,* 515, (February), 31–36, 2022, doi: 10.1016/j.nimb.2022.01.012.

[57] D. Brunner & V. Glebovsky, Analysis of flow-stress measurements of high-purity tungsten single crystals, *Mater. Lett.*, 44, (3), 144–152, 2000, doi: 10.1016/S0167-577X(00)00017-3.

[58] F. Chen, Z. Cao, G. Chen, X. Kai, J. Wu, & Y. Zhao, Synchrotron radiation micro-beam analysis of the effect of strontium on primary silicon in Zn–27Al–3Si alloy, *J. Alloys Compd.*, 749, 575–579, 2018, doi: 10.1016/j.jallcom.2018.03.256.

[59]  K. Sakamoto, F. Hayashi, K. Sato, M. Hirano, & N. Ohtsu, XPS spectral analysis for a multiple oxide comprising NiO, TiO2, and NiTiO3, *Appl. Surf. Sci.*, 526, (April), 2020, doi: 10.1016/j.apsusc.2020.146729.

[60]  N. E. Gorji, R. O'Connor, & D. Brabazon, X-ray tomography, AFM and nanoindentation measurements for recyclability analysis of 316L powders in 3D printing process, *Procedia Manuf.*, 47, 1113–1116, 2020, doi: 10.1016/j.promfg.2020.04.127.

[61]  P. Eppmann, B. Prüger, & J. Gimsa, Particle characterization by AC electrokinetic phenomena 2. Dielectrophoresis of Latex particles measured by dielectrophoretic phase analysis light scattering (DPALS), *Colloids Surfaces A Physicochem. Eng. Asp.*, 149, (1–3), 443–449, 1999, doi: 10.1016/S0927-7757(98)00304-5.

[62]  B. Dubiel, D. Wolf, & A. Czyrska-Filemonowicz, TEM and electron holography analyses of granular and thin layered Cu-Co magnetic materials, *Ultramicroscopy*, 110, (5), 433–437, 2010, doi: 10.1016/j.ultramic.2009.10.017.

[63]  K. V. Durga Rajesh, A. V. S. Ram Prasad, A. Munaf Shaik, & T. Buddi, SEM with EDAX analysis on plasma arc welded butt joints of AISI 304 and AISI 316 steels, *Mater. Today Proc.*, 44, 1350–1355, 2021, doi: 10.1016/j.matpr.2020.11.393.

[64]  S. Derrough, A. Pikon, C. Sourd, I. Guérin, J. Simon, S. Gaugiran, T. Cardolaccia, Y. Liu, P. Trefonas, G. Barclay, & Y. C. Bae, Thermal characterization of materials for double patterning, *Microelectron. Eng.*, 87, (5–8), 997–1000, 2010, doi: 10.1016/j.mee.2009.11.117.

[65]  M. Toto Jamil, A. Benabou, S. Clénet, S. Shihab, L. Le Bellu Arbenz, & J. C. Mipo, Magneto-thermal characterization of bulk forged magnetic steel used in claw pole machine, *J. Magn. Magn. Mater.*, 502, (January), 166526, 2020, doi: 10.1016/j.jmmm.2020.166526.

[66]  K. Korhammer, M. M. Druske, A. Fopah-Lele, H. U. Rammelberg, N. Wegscheider, O. Opel, T. Osterland, & W. Ruck, Sorption and thermal characterization of composite materials based on chlorides for thermal energy storage, *Appl. Energy*, 162, 1462–1472, 2016, doi: 10.1016/j.apenergy.2015.08.037.

[67]  N. Yu, R. Z. Wang, Z. S. Lu, & L. W. Wang, Development and characterization of silica gel-LiCl composite sorbents for thermal energy storage, *Chem. Eng. Sci.*, 111, 73–84, 2014, doi: 10.1016/j.ces.2014.02.012.

[68]  E. P. Del Barrio, J. L. Dauvergne, & C. Pradere, Thermal characterization of materials using Karhunen-Loève decomposition techniques – Part I. Orthotropic materials, *Inverse Probl. Sci. Eng.*, 20, (8), 1115–1143, 2012, doi: 10.1080/17415977.2012.658388.

[69]  E. P. Del Barrio, J. L. Dauvergne, & C. Pradere, Thermal characterization of materials using Karhunen-Loève decomposition techniques – Part II. Heterogeneous materials, *Inverse Probl. Sci. Eng.*, 20, (8), 1145–1174, 2012, doi: 10.1080/17415977.2012.658517.

[70]  E. Borri, J. Y. Sze, A. Tafone, A. Romagnoli, Y. Li, & G. Comodi, An experimental and numerical method for thermal characterization of phase change materials for cold thermal energy storage, *Energy Procedia*, 158, 5041–5046, 2019, doi: 10.1016/j.egypro.2019.01.657.

[71]  S. Søgaard, M. Bryder, M. Allesø, & J. Rantanen, Characterization of powder properties using a powder rheometer, *Proceedings of Electronic Conference on Pharmaceutical Sciences* , 2, 1–8, 2012, https://sciforum.net/manuscripts/825/man uscript.pdf.

[72]  L. A. Mills & I. C. Sinka, Effect of particle size and density on the die fill of powders., *European J. Pharm. Biopharm.*, 84, 642–652, 2013.

[73]  W. Nan, M. Ghadiri, & Y. Wang, Analysis of powder rheometry of FT4 – Effect of particle shape, *Chem. Eng. Sci.*, 173, 374–383, 2017.

[74]  T. Novoselova, S. Celotto, R. Morgan, P. Fox, & W. O'Neill, Formation of TiAl intermetallics by heat treatment of cold-sprayed precursor deposits, *J. Alloys Compd.*, 436, (1–2), 69–77, 2007, doi: 10.1016/j.jallcom.2006.06.101.

[75]  A. Lawley, Physicochemical considerations in powder metallurgy, *JOM*, 42, 12–14, 1990, doi.org/10.1007/BF03220916.

[76]  O. Yeheskel & M. P. Dariel, The effect of processing on the elastic moduli of porous γ-TiAl, *Mater. Sci. Eng. A*, 354, (1–2), 344–350, 2003, doi: 10.1016/S0921-5093(03)00037-6.

[77]  T. Maurizi Enrici, O. Dedry, F. Boschini, J. T. Tchuindjang, & A. Mertens, Microstructural and thermal characterization of 316L + WC composite coatings obtained by laser cladding, *Adv. Eng. Mater.*, 22, (12), 1–12, 2020, doi: 10.1002/adem.202000291.

[78]  M. Trejo-Duran, A. Martınez-Richa, R. Vera-Graziano, & 1 V. M. Castan˜o Alvarado-Me´ndez, Structural and thermal characterization of hybrid materials based on TEOS and DCN, *Wiley Intersci.*, (5), 1–11, 2008, doi: 10.1002/app.29110.

[79]  S. M. Pérez-Moreno, M. J. Gázquez, A. G. Barneto, & J. P. Bolívar, Thermal characterization of new fire-insulating materials from industrial inorganic TiO2 wastes, *Thermochim. Acta*, 552, 114–122, 2013, doi: 10.1016/j.tca.2012.10.021.

[80]  Y. N. Zhang, R. Z. Wang, Y. J. Zhao, T. X. Li, S. B. Riffat, & N. M. Wajid, Development and thermochemical characterizations of vermiculite/SrBr2 composite sorbents for low-temperature heat storage, *Energy*, 115, 120–128, 2016, doi: 10.1016/j.energy.2016.08.108.

[81]  X. Zhao, S. A. Mofid, M. R. A. Hulayel, G. W. Saxe, B. P. Jelle, & R. Yang, Reduced-scale hot box method for thermal characterization of window insulation materials, *Appl. Therm. Eng.*, 160, (June), 114026, 2019, doi: 10.1016/j.applthermaleng.2019.114026.

[82]  J. Pelleg, *Mechanical Properties of Materials. Solid Mechanics and Its Applications,*. New York: Springer Dordrecht Heidelberg, 2013. doi: 10.1007/978-94-007-4342-7.

[83]  G. E. Dieter, *Mechanical Metallurgy*. McGraw Hill, 1961.

[84]  S. S. Murugan, Mechanical properties of materials: Definition, testing and application, *Int. J. Mod. Stud. Mech. Eng.*, 6, (2), 28–38, 2020, doi: 10.20431/2454-9711.0602003.

[85]  L. Tshabalala, O. Sono, W. Makoana, J. Masindi, O. Maluleke, C. Johnston, & S. Masete, Axial fatigue behaviour of additively manufactured tool steels, *Mater. Today Proc.*, 38, 789–792, 2021, doi: 10.1016/j.matpr.2020.04.548.

[86]  G. Chitos, Understanding the different types of hardness tests, *WillRich Precision Instrument*, 2022. https://willrich.com/understanding-different-types-hardness-tests

[87]  F. P. Beer, E. R. Johnston, J. T. DeWolf, & D. F. Mazurek, *Mechanics of Materials*, fifth edition. McGraw Hill, 2009.

[88]  T. Inamura, Y. Takahashi, H. Hosoda, K. Wakashima, T. Nagase, T. Nakano, Y. Umakoshi, & S. Miyazaki, Martensitic transformation behavior and shape memory properties of Ti–Ni–Pt melt-spun ribbons, *Mater. Trans.*, 47(3) 540–545, 2006.

[89]  Y. Yamabe-Mitarai, R. Arockiakumar, A. Wadood, K. S. Suresh, T. Kitashima, T. Hara, M. Shimojo, W. Tasaki, M. Takahashi, S. Takahashi, & H. Hosoda, Ti(Pt, Pd, Au) based high temperature shape memory alloys, *Mater. Today Proc.*, 2, S517–S522, 2015.

[90]  T. Biggs, M. B. Cortie, M. J. Witcomb, & L. A. Cornish, Platinum alloys for shape memory applications, *Platin. Met. Rev.*, 47, (4), 142–156, 2003.

[91]  G. Lütjering, J. C. Williams, & A. Gysler, Microstructure and mechanical properties of titanium alloys, *Microstruct. Prop. Mater.*, 2, 1–74, 2000.

[92]  J. Preussner, M. Wenderoth, S. Prins, R. Völkl, & U. Glatzel, Platinum alloy development – the Pt Al-Cr-Ni system, International Platinum Conference 'Platinum Surges Ahead', The Southern African Institute of Mining and Metallurgy, 6, 103–106, 2006.

[93]  Y. Pan, Y. Lin, X. Wang, S. Chen, L. Wang, C. Tong, & Z. Cao, Structural stability and mechanical properties of Pt-Zr alloys from first-principles, *J. Alloys Compd.*, 643, 49–55, 2015.

[94]  A. Wadood, M. Takahashi, S. Takahashi, H. Hosoda, & Y. Yamabe-Mitarai, High-temperature mechanical and shape memory properties of TiPt-Zr and TiPt-Ru alloys, *Mater. Sci. Eng. A*, 564, 34–41, 2013.

[95]  T. Fukuta, M. Iiba, Y. Kitagawa, & Y. Sakai, Experimental study on stress-strain property of shape memory alloy and its application to self-restoration of structural members, *13th Wordl Conf. Earthq. Eng.*, 610, 1–9, 2004.

[96]  J. Mohd Jani, M. Leary, A. Subic, & M. A. Gibson, A review of shape memory alloy research, applications and opportunities, *Mater. Des.*, 56, 1078–1113, 2014.

[97]  T. Biggs, M. B. Cortie, M. J. Witcomb, & L. A. Cornish, Martensitic transformations, microstructure, and mechanical workability of TiPt, *Metall. Mater. Trans. A*, 32, (August), 1881–1886, 2001.

# Index

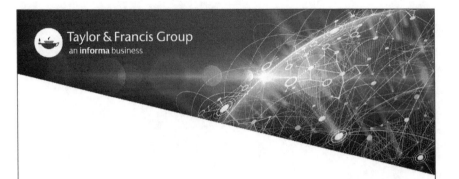